KS1 Success

Age 5-7

Maths

Practice Test Papers

Alan Dobbs

SATs

Practice Papers

Contents

Introduction and instructions

How these tests will help your child

This book is made up of three complete sets of practice test papers. Each set contains similar test papers to those that your child will take in maths at the end of Year 2. The tests will assess your child's knowledge, skills and understanding in the areas of study undertaken since they began Year 1. These practice test papers can be used any time throughout the year to provide practice for the Key Stage 1 tests.

The results of each set of papers will provide a good idea of the strengths and weaknesses of your child.

Administering the tests

- Provide your child with a quiet environment where they can complete each test undisturbed.
- Provide your child with a pen or pencil and an eraser. A ruler is allowed for Paper 2 but a calculator is **not** allowed in either paper.
- The amount of time given for each test varies, so remind your child at the start of each one how long they have and give them access to a clock or watch.
- You should only read the instructions out to your child, not the actual questions (except the aural questions).
- Although handwriting is not assessed, remind your child that their answers should be clear.
- Advise your child that if they are unable to do one of the questions, they should go on to the next one and come back to it later, if they have time. If they finish before the end, they should go back and check their work.

Paper 1: arithmetic

- All answers are worth 1 mark, with a total number of 25 marks for each test.
- Your child will have approximately **20 minutes** to complete the test.
- The questions test your child's understanding and application of number, calculations and fractions.
- Both formal and informal methods of answering mathematical problems are acceptable.

Paper 2: reasoning

- All answers are worth 1 mark (unless stated), with a total number of 35 marks for each test.
- Your child will have approximately **35 minutes** to complete the test, including 5 minutes for the aural questions.
- The first five questions should be read out to your child. Using the aural questions administration guide on pages 95–96, read each question and remember to repeat the bold text.
- Questions will test calculation, data gathering and interpretation, shape, space and measures. They will test your child's ability to use maths in contextualised, abstract and real-life situations.

Marking the practice test papers

The answers have been provided to enable you to check how your child has performed. Fill in the marks that your child achieved for each part of the tests.

Please note: these tests are **only a guide** to the mark your child can achieve and cannot guarantee the same is achieved during the Key Stage 1 tests, as the mark needed to achieve the expected standard varies from year to year.

	Set A	Set B	Set C
Paper 1: arithmetic	/25	/25	/25
Paper 2: reasoning	/35	/35	/35
Total	/60	/60	/60

The scores achieved on these practice test papers will help to show if your child is working at the expected standard in maths:
20–30 = working towards the expected standard
31–42 = working at the expected standard
43–60 = working above the expected standard.

When an area of weakness has been identified, it is useful to go over these, and similar types of questions, with your child. Sometimes your child will be familiar with the subject matter but might not understand what the question is asking. This will become apparent when talking to your child.

Shared marking and target setting

Engaging your child in the marking process will help them to develop a greater understanding of the tests and, more importantly, provide them with some ownership of their learning. They will be able to see more clearly how and why certain areas have been identified for them to target for improvement.

Top tips for your child

Don't make silly mistakes. Make sure you emphasise to your child the importance of reading the question. Easy marks can be picked up by just doing as the question asks.

Make answers clearly legible. If your child has made a mistake, encourage them to put a cross through it and write the correct answer clearly next to it. Try to encourage your child to use an eraser as little as possible.

Don't panic! These practice test papers, and indeed the end of Key Stage 1 tests, are meant to provide a guide to the standard a child has attained. They are not the be-all and end-all, as children are assessed regularly throughout the school year. Explain to your child that there is no need to worry if they cannot do a question – tell them to go on to the next question and come back to the problematic question later if they have time.

Key Stage 1

SET A

Maths

PAPER 1

Maths

Paper 1: arithmetic

Time:

You have approximately **20 minutes** to complete this test paper.

Maximum mark	Actual mark
25	

First name	
Last name	

Practice question

$5 - 1 =$ ☐

1

$5 + 3 =$ ☐

2

$10 - 3 =$ ☐

3

$10 \div 2 =$ ☐

4

$12 + \square = 19$

5

$10 + 11 = \square$

6

$\frac{1}{4}$ of $20 = \square$

7

$5 \times 2 = 12 - \square$

8

$\square - 16 = 16$

9

$5 \times 3 = \square$

10

$\frac{3}{4}$ of 20 = ☐

11

99 = 59 + ☐

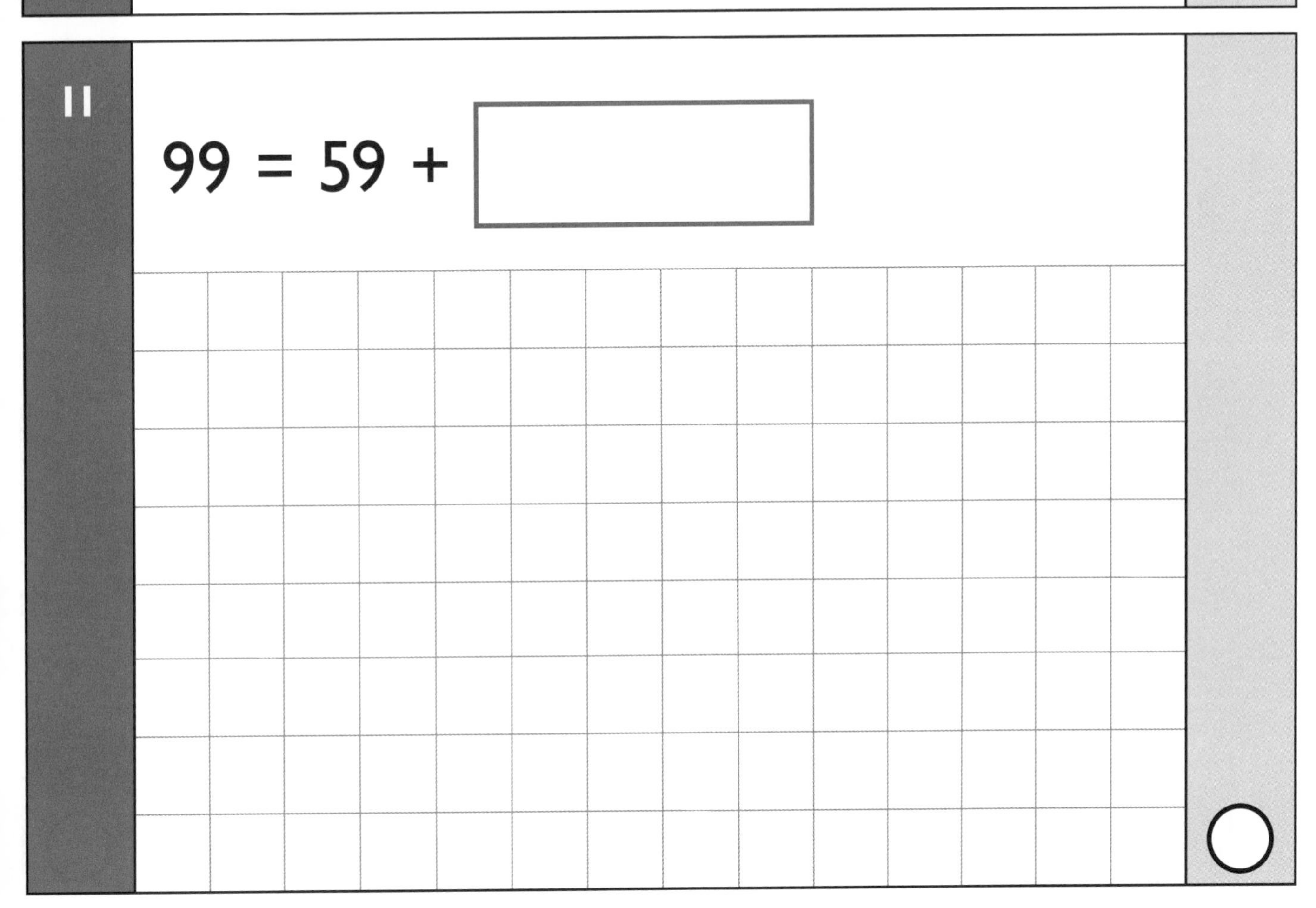

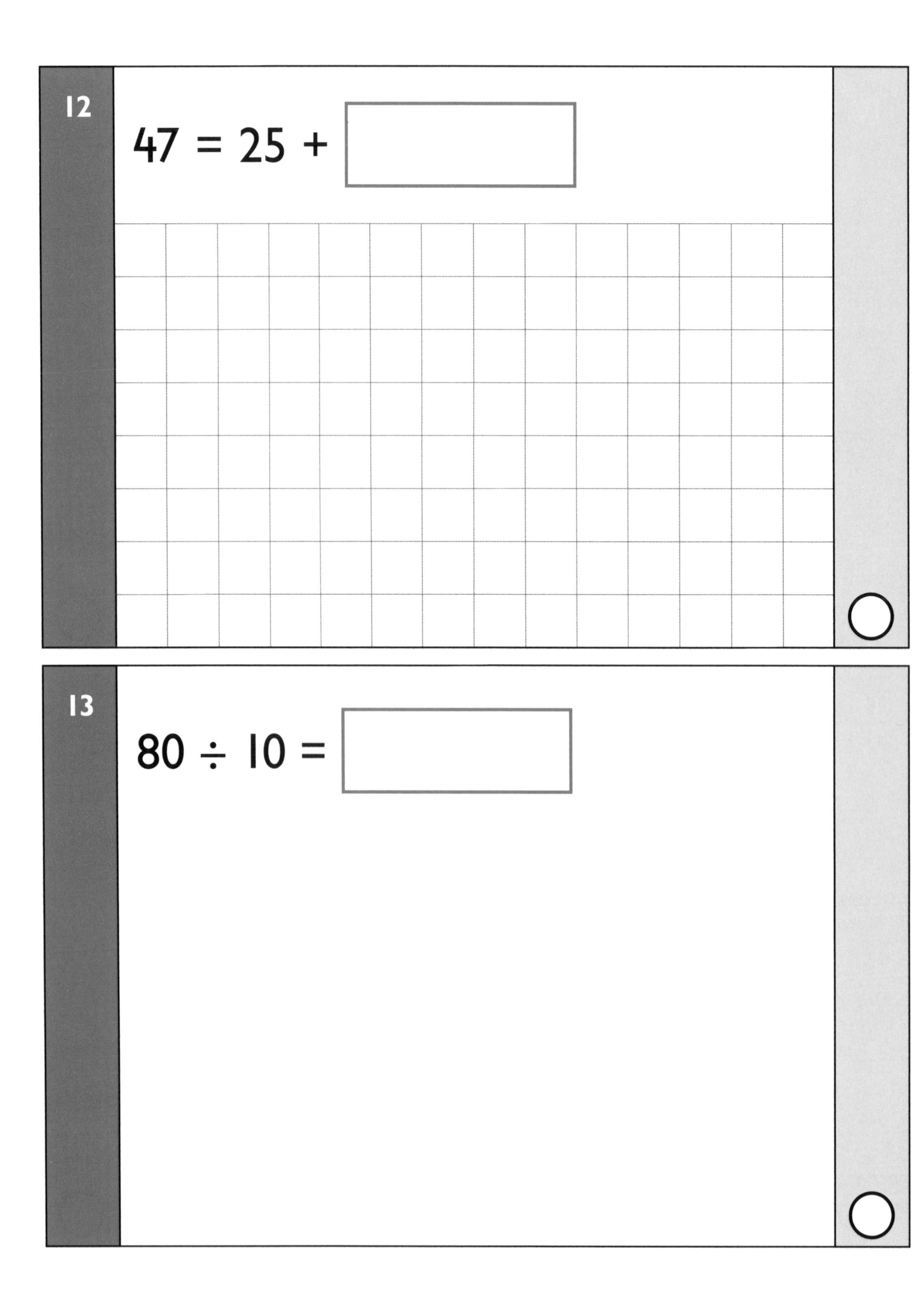
12
47 = 25 +
13
80 ÷ 10 =

14

$\square - 14 = 30$

15

$25 + 16 = 41$

$41 - \square = 16$

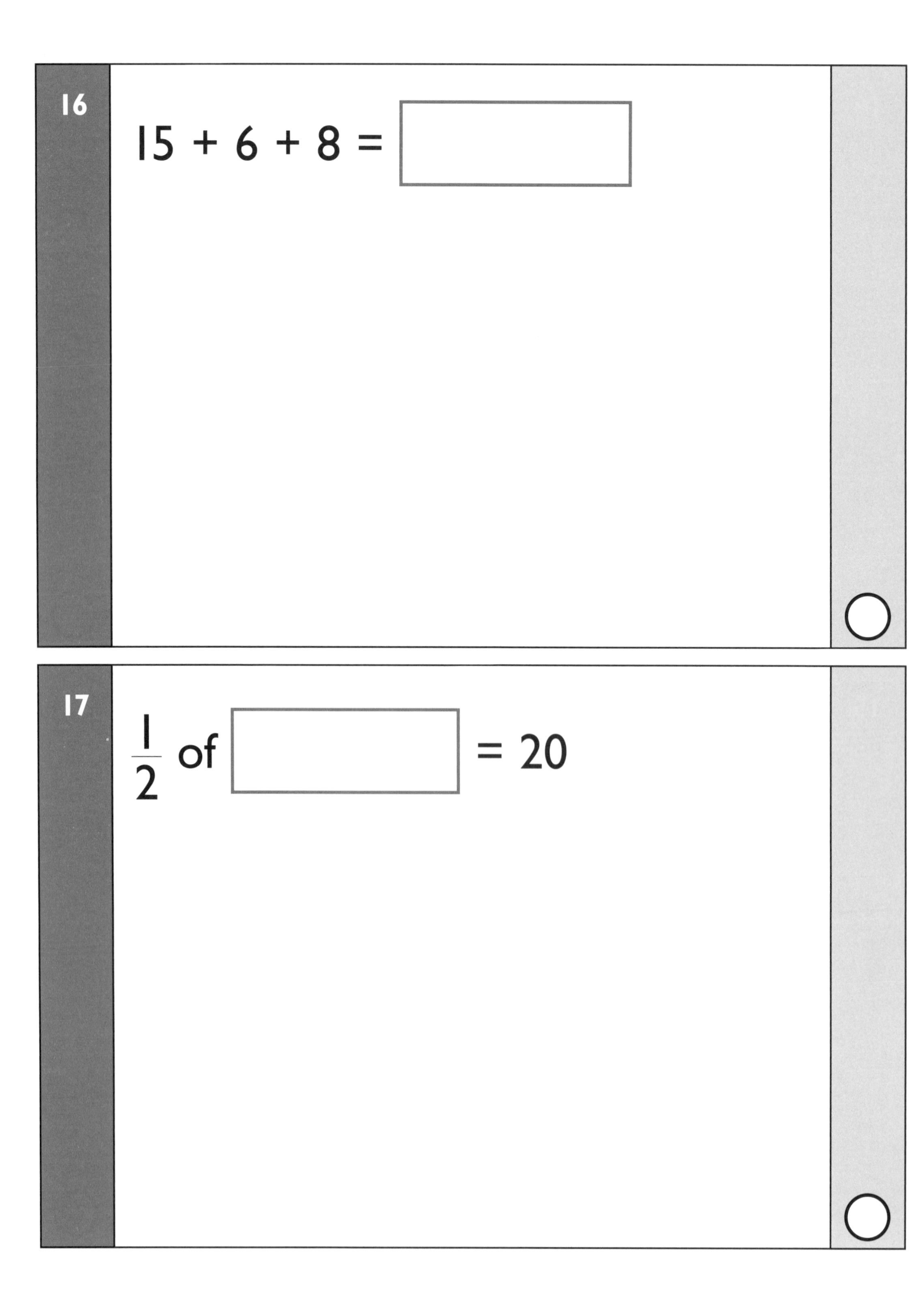
16
15 + 6 + 8 =
17
$\frac{1}{2}$ of
= 20

18

$20 \div 2 = 2 \times$ ☐

19

$18 = \frac{1}{2}$ of ☐

20

$30 \div \square = 2 \times 5$

21

$20 \times 2 = \square$

22

$\frac{1}{3}$ of 21 = ☐

23

45 + 13 + 2 = ☐

24

$\frac{2}{4}$ of 10 = ☐

25

12 = $\frac{1}{2}$ of ☐

Key Stage 1

Maths

Paper 2: reasoning

You will need to ask someone to read the instructions to you for the first five questions. These can be found on page 95. You can answer the remaining questions on your own.

Time:

You have approximately **35 minutes** to complete this test paper. This timing includes approximately 5 minutes for the aural questions.

Maximum mark	Actual mark
35	

First name	
Last name	

Practice question

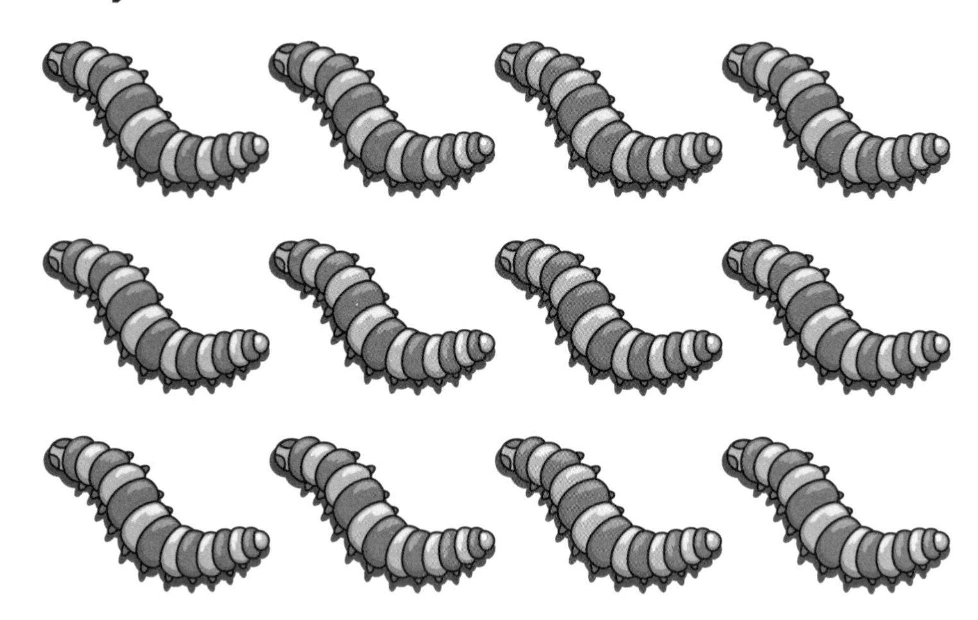

1

2

stickers

3

100 g **3 kg** **20 kg** **100 kg**

4

p

5

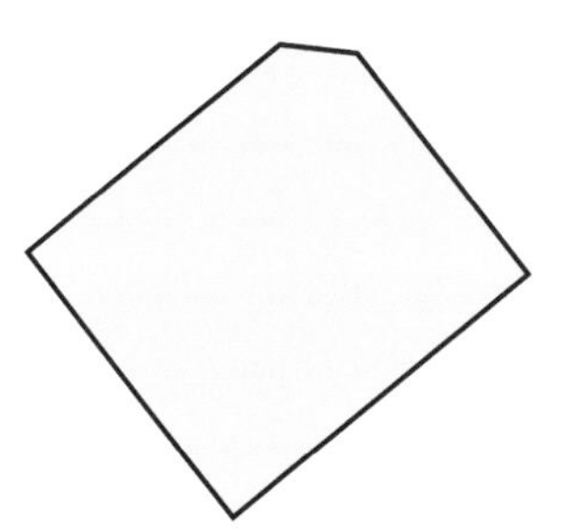

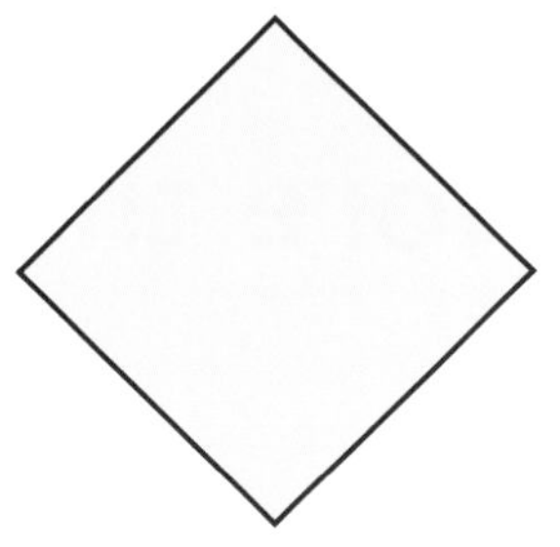

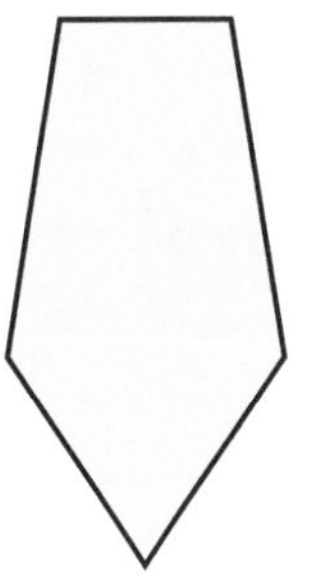

Now continue with the rest of the paper on your own.

6 Look at these signs.

+	=	–

Write a sign in each box to make this correct.

23 ☐ 8 ☐ 15

7 This sentence is correct.

15 is more than **10** ✓

Four of these sentences are correct.

Tick (✓) them.

45 is less than **57** ☐

34 is more than **27** ☐

69 is more than **96** ☐

81 is equal to **18** ☐

28 is less than **34** ☐

59 is more than **7** ☐

8

There are **26** children.

8 children are drawing.

How many children are **not drawing**?

[ ] children

9 Some of these numbers are **even**.

Circle the **even** numbers.

13 16 29 33 76 50 27

10 Complete this table.

The first row has been done for you.

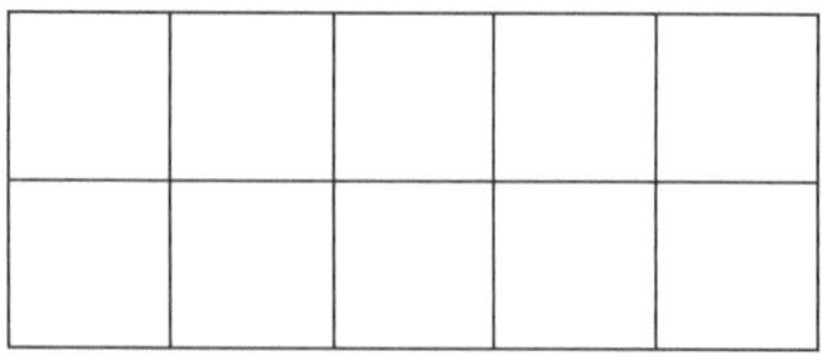	2 × 5	10
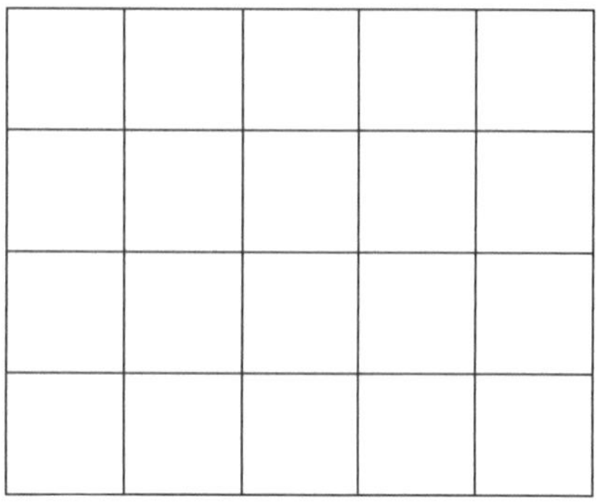		20

11 Put these numbers in order from **least** to **most**.

14 98 37 78 100 29 16

least most

12 Circle $\frac{3}{4}$ of the bananas.

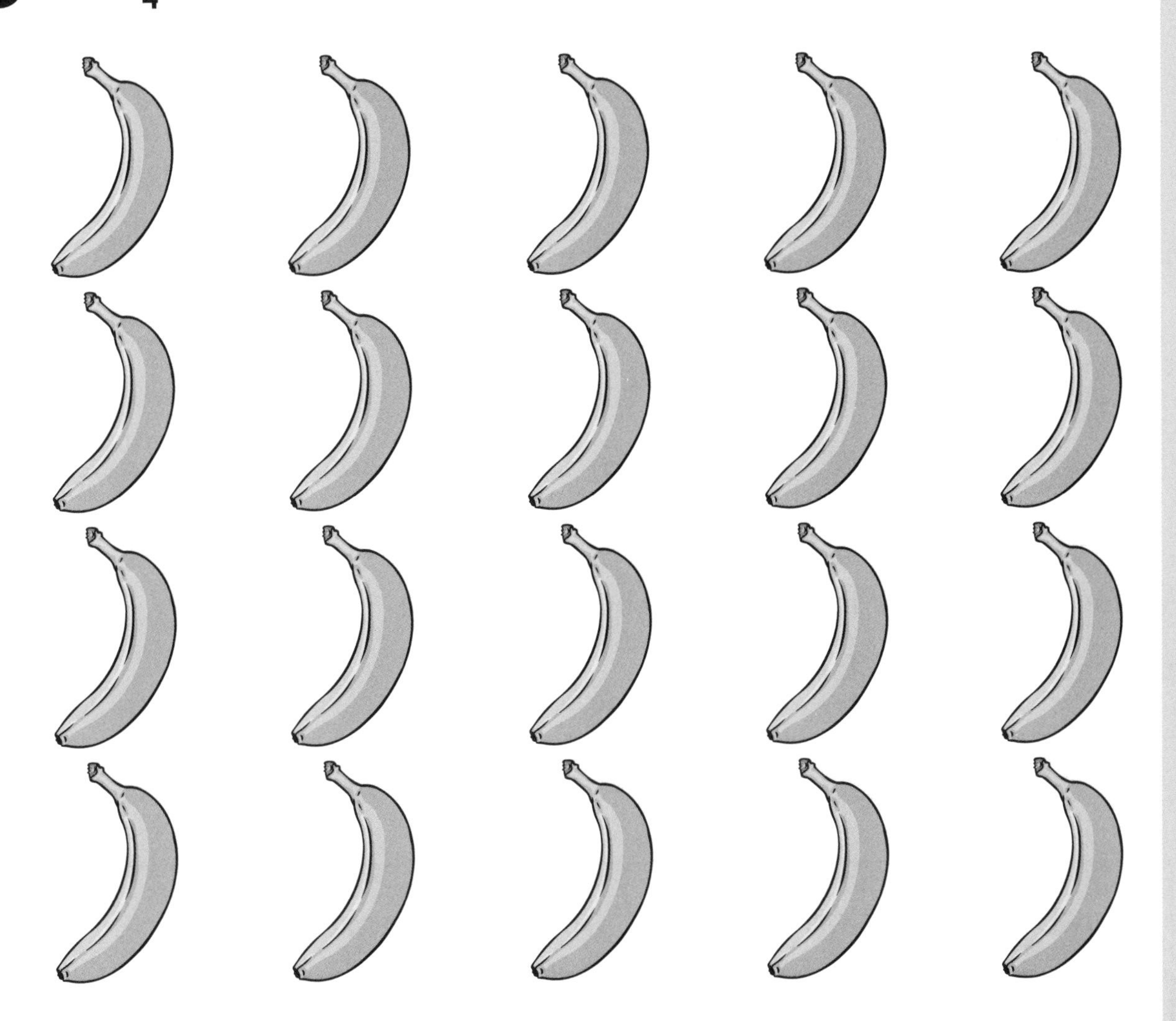

13 Shade this jug to show 650 ml.

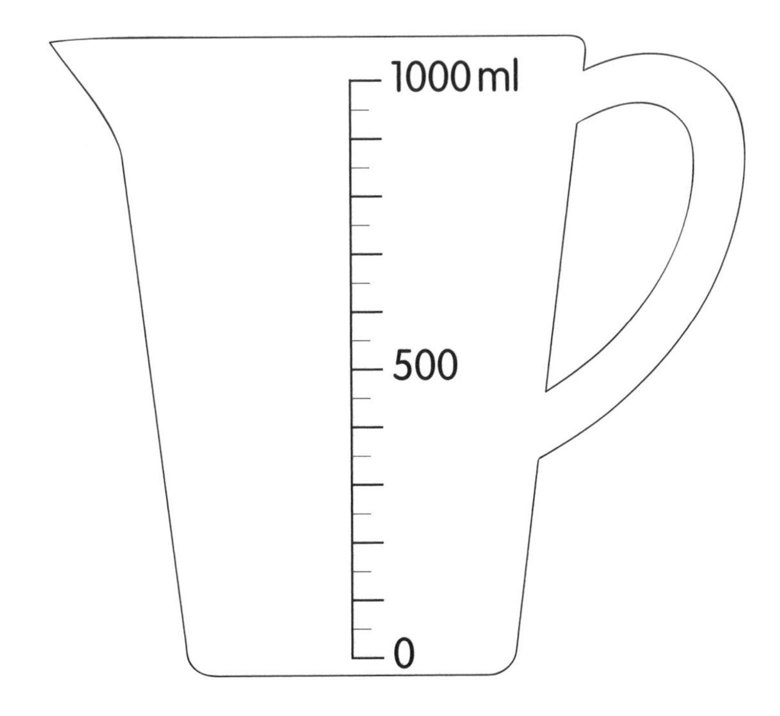

14

How many 10p coins are there in £1.50?

10p coins

15 Write each number in the correct box.

One has been done for you.

51 28 16 43 32 34

Numbers less than 30	Numbers more than 30
16	

16 Look at these cards.

Use each card once to write each of these number sentences.

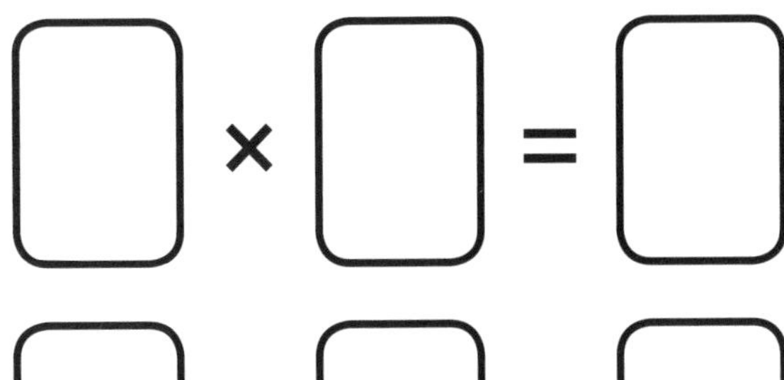

17 Tick (✓) the two shapes that are hexagons.

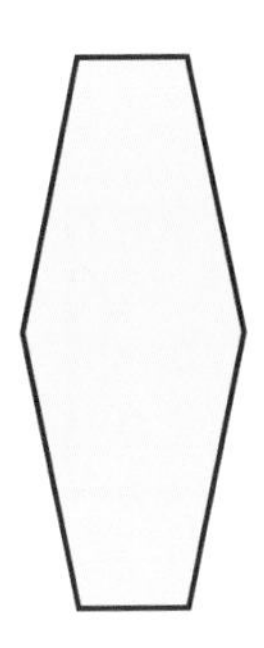

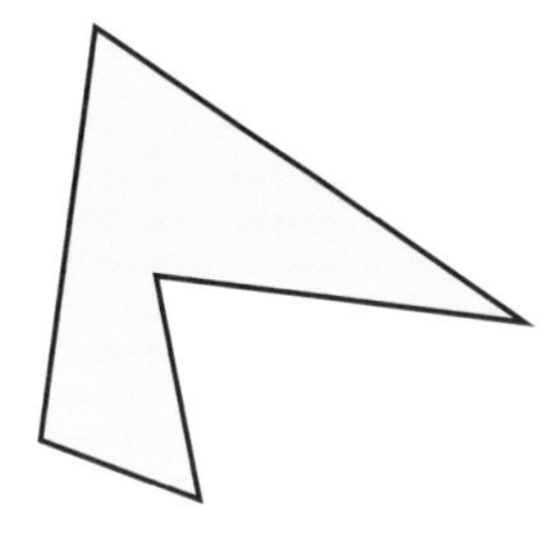

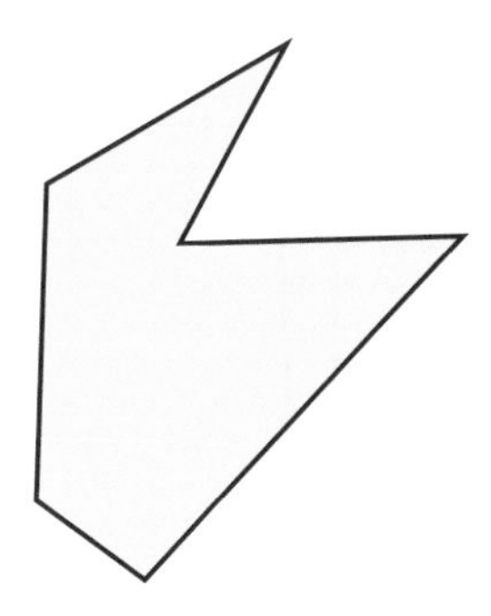

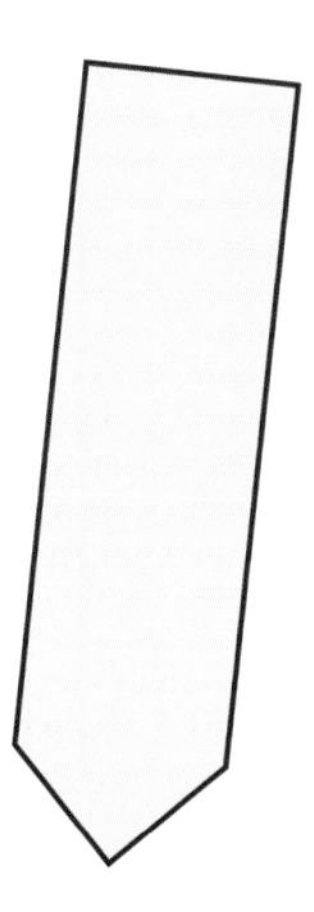

18 Pineapples cost 50p each. Oranges cost 10p each.

Adam buys **2** pineapples and **3** oranges.

How much change does he get from £1.50?

Show your working

p

2 marks

19 There are **100** pieces in a kit.

15 pieces have been lost.

How many pieces remain?

pieces

20 Class 5 have made a bar chart.

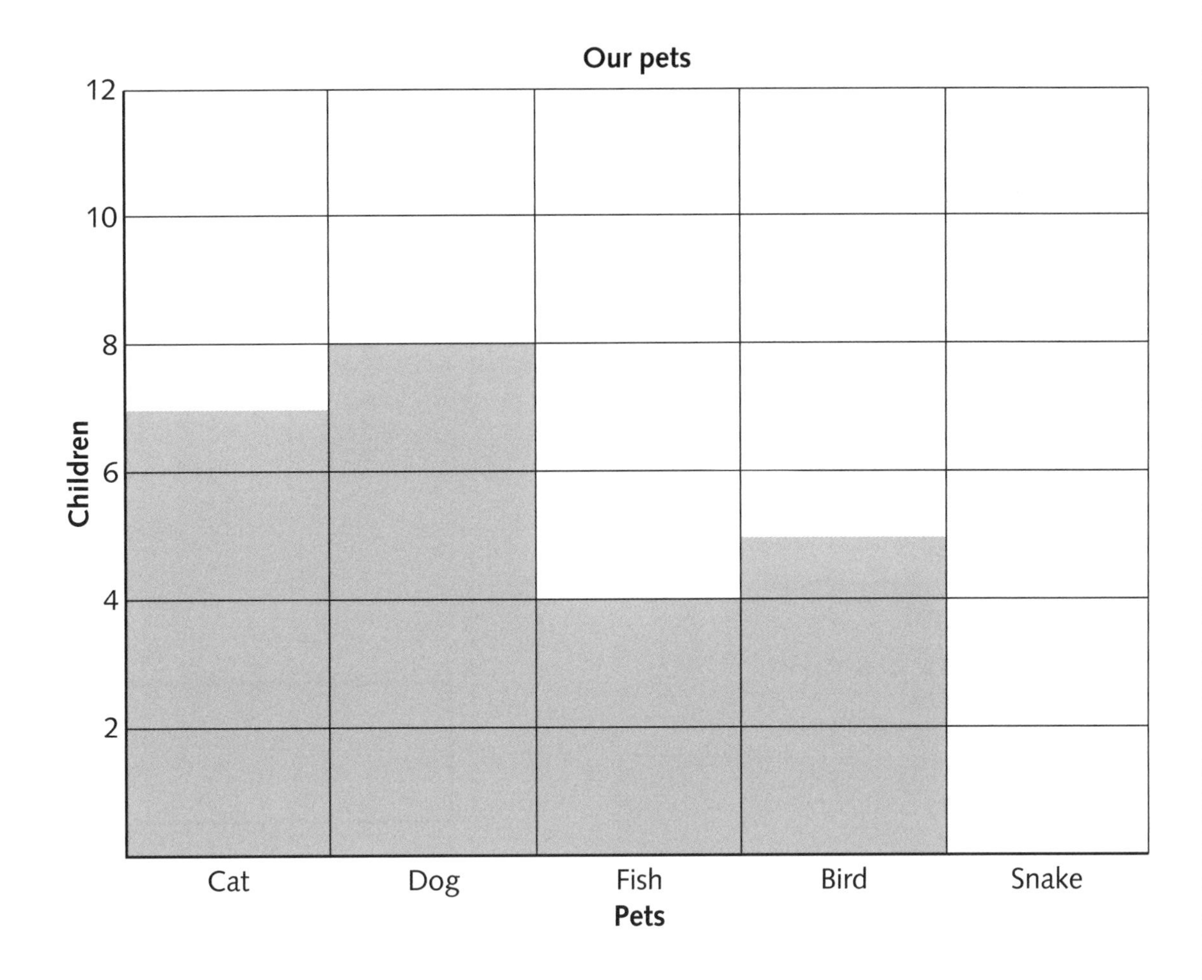

a) 3 children have a snake.

Show this on the bar chart.

b) More children have a cat than a fish.

How many more?

[ ] children

21 Complete the sequences.

a)

2	5		11		17

b)

24			15		9

22 Look at the thermometer.

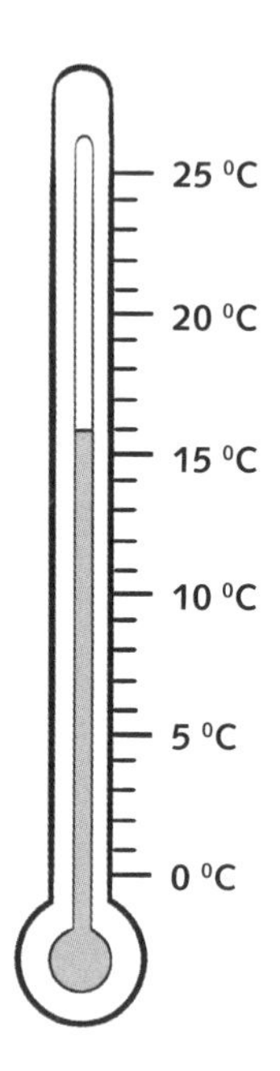

What temperature is shown?

°C

23

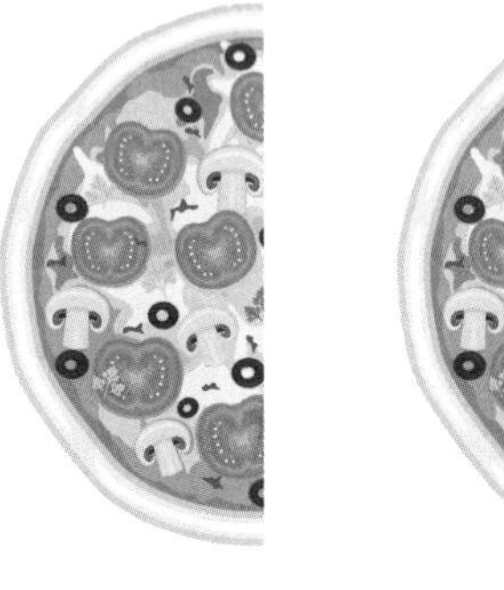
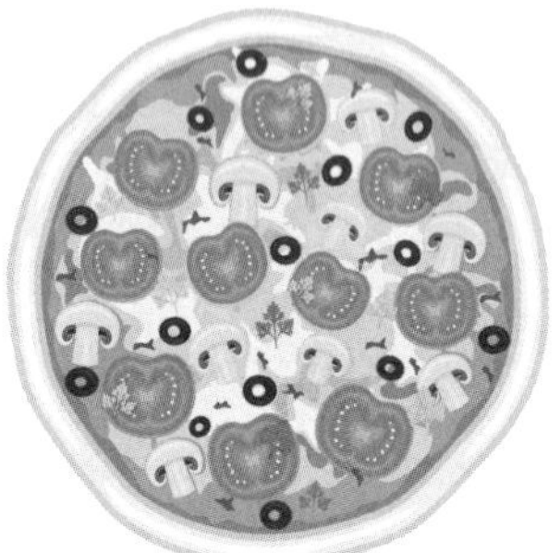
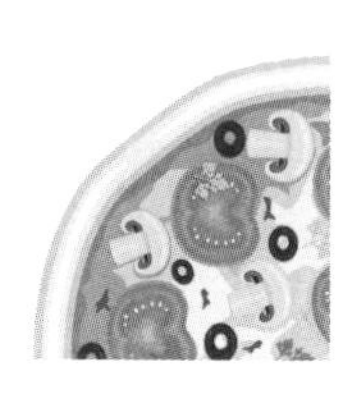
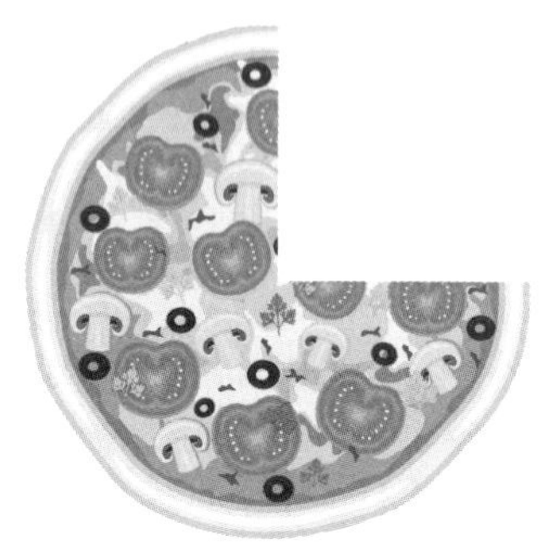

1 whole	$\frac{1}{4}$	$\frac{3}{4}$	$\frac{1}{2}$

a) Join the fractions to the correct pizza.

b) Add the total amount of pizza.

How many **whole** pizzas?

whole pizzas

24 Draw the hands on the analogue (dial) clocks so that they show the following times.

half past 8

quarter past 6

5 o'clock

25 Estimate the number marked by the arrow.

Write the number in the box.

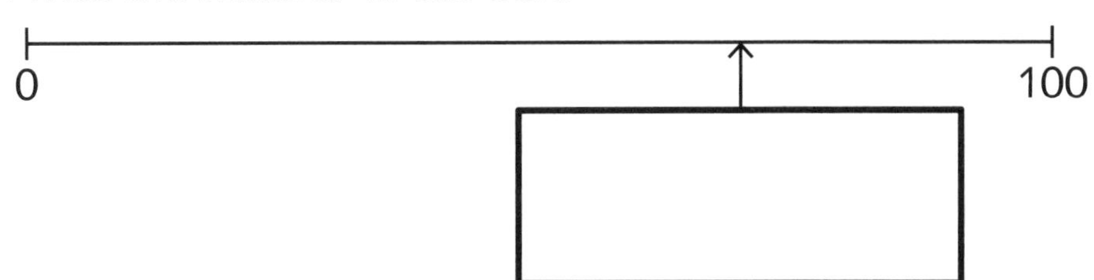

26 Look at this arrow.

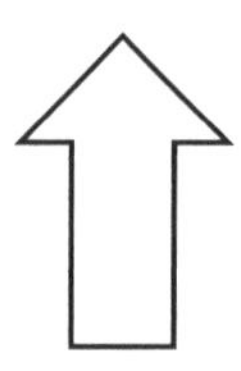

You turn the arrow one clockwise quarter turn.

Tick (✓) the arrow that shows how it looks after the turn.

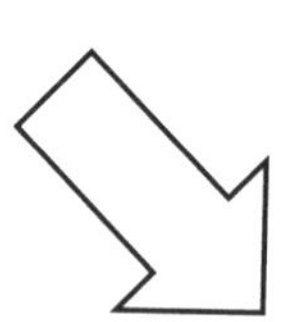

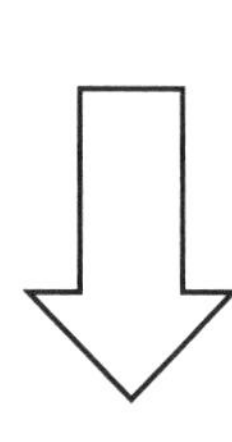

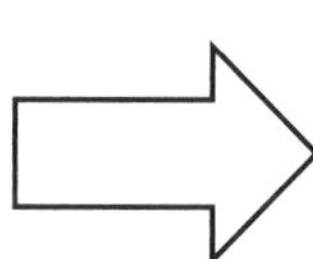

27 Match the correct name with each 3-D shape.

Cube

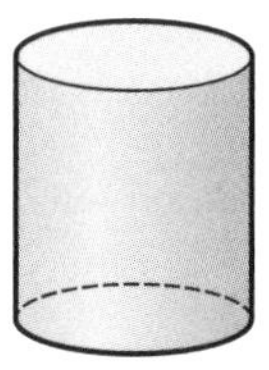

Cylinder

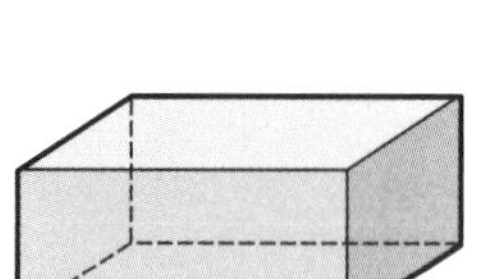

Pyramid

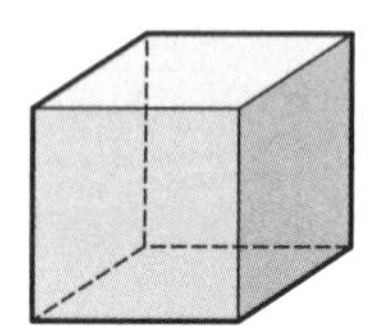

Cuboid

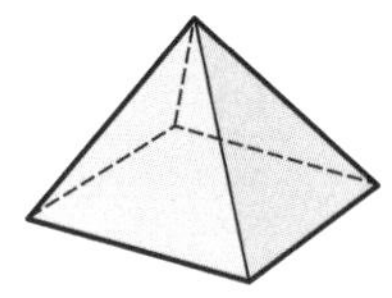

28 This reel has 1 m of tape.

Hamid needs to cut the tape into 20 cm pieces.

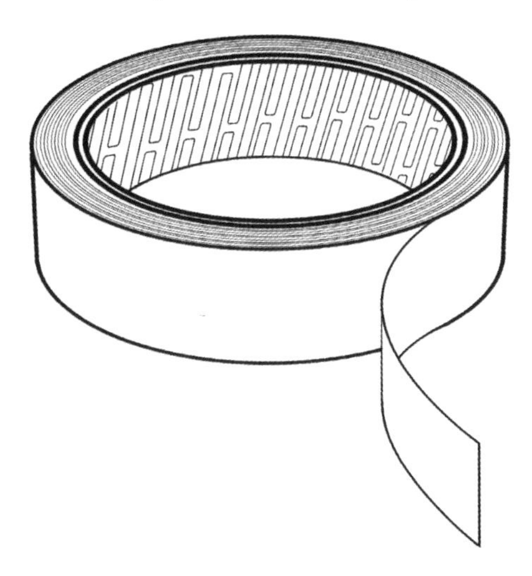

How many pieces will he have if he uses all of the tape?

[ ] pieces

29

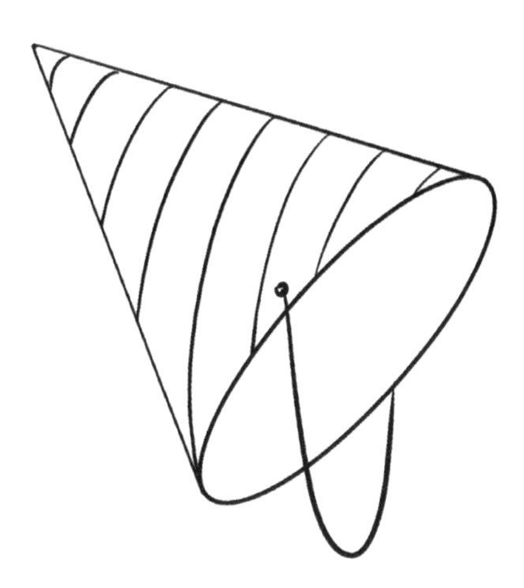

Party hats are sold in packs of 5.

Jane needs 27 hats for her party.

How many packs of hats should she buy?

[ ] packs

30 Mark the vertices that you can see on this cube by circling them.

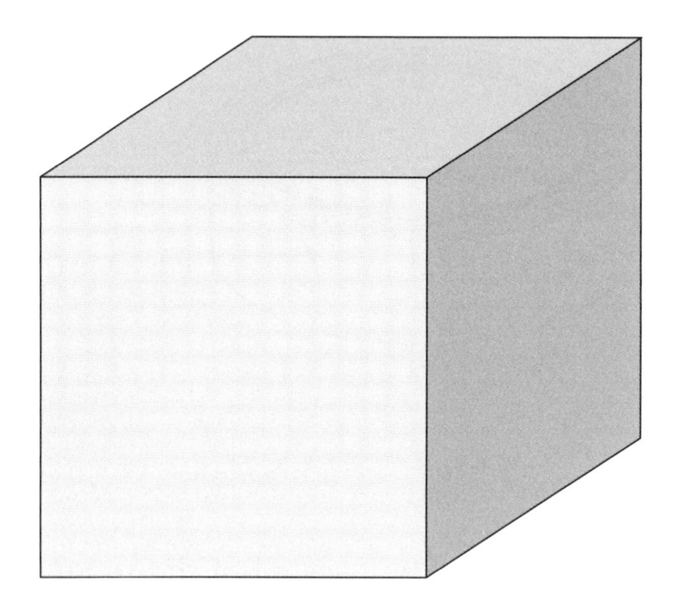

31 Circle the bee that is 6th from the left.

Key Stage 1

SET
B

Maths

PAPER 1

Maths

Paper 1: arithmetic

Time:

You have approximately **20 minutes** to complete this test paper.

Maximum mark	Actual mark
25	

First name	
Last name	

Practice question

5 + 7 = ☐

1

$99 + 1 =$ ☐

2

$30 \div 2 =$ ☐

3

$11 + 3 + 5 =$ ☐

4

[ ] $= \frac{1}{4}$ of 40

5

44 − [ ] = 30

6

$\frac{3}{4}$ of 12 = [ ]

7

$40 + \square = 100$

$100 = 60 + \square$

8

$14 - \square = 7$

9

$10 \times 10 = \square$

10

$\frac{1}{3}$ of 30 = ☐

11

56 – 31 = ☐

12

$15 + 7 + 9 = \square$

13

$15 + \square = 35$

14

Half of 20 = ☐

15

$25 \div 5 =$ ☐

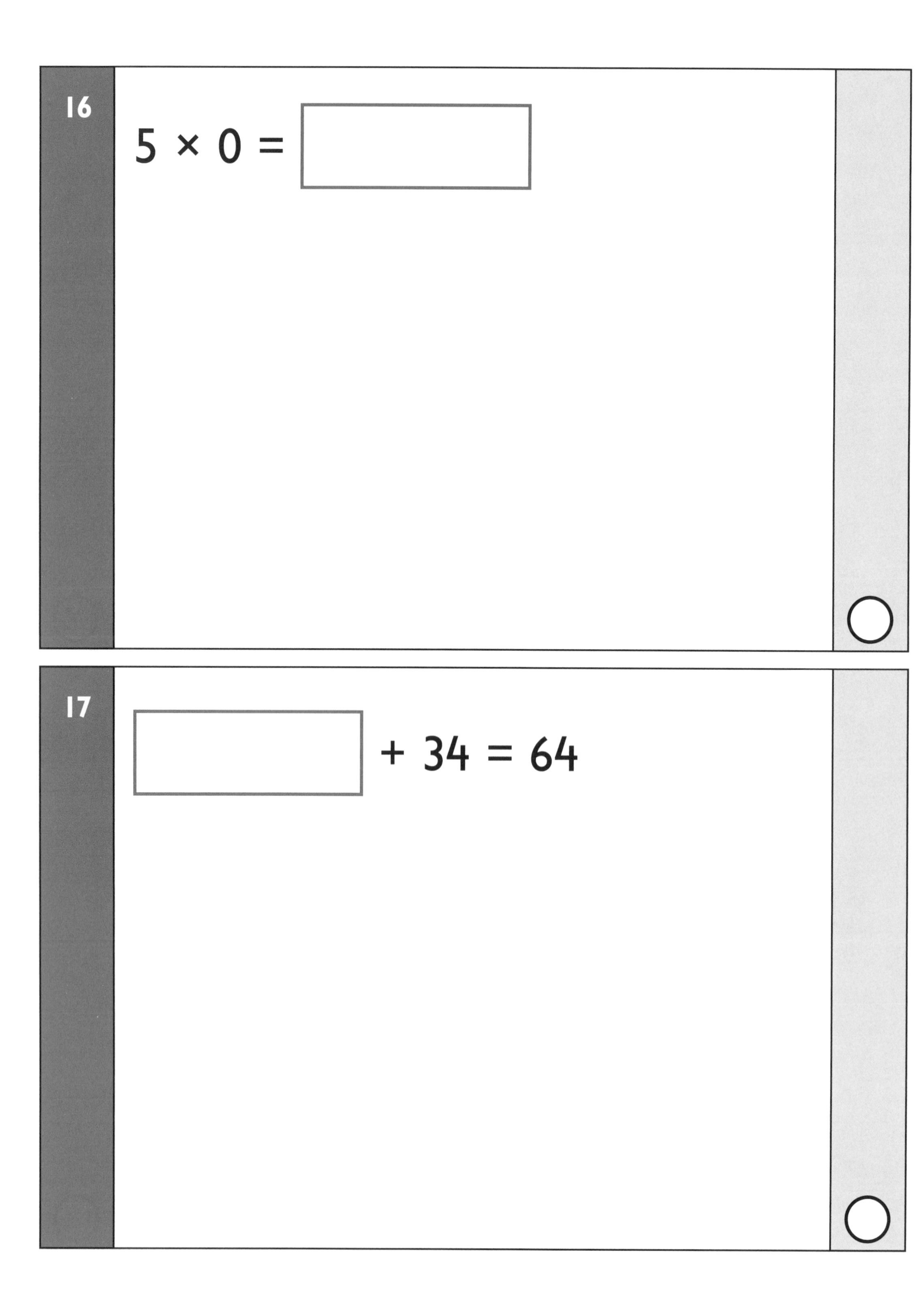
16
5 × 0 =
17
+ 34 = 64

18

$25 - 6 - 2 = \square$

19

$\frac{3}{4}$ of $100 = \square$

20

$5 + 5 + 20 = \square$

21

$67 - \square = 25$

22

$25 + \square = 50$

23

$100 - \square = 60$

24

$\frac{1}{3}$ of 66 = ☐

25

50 × ☐ = 100

Key Stage 1

SET
B

Maths

PAPER 2

Maths

Paper 2: reasoning

You will need to ask someone to read the instructions to you for the first five questions. These can be found on page 96. You can answer the remaining questions on your own.

Time:

You have approximately **35 minutes** to complete this test paper. This timing includes approximately 5 minutes for the aural questions.

Maximum mark	Actual mark
35	

First name	
Last name	

Practice question

stickers

1

2

3

4

faces

5

10 minutes **30 minutes** **2 hours** **12 hours**

Now continue with the rest of the paper on your own.

6 Tara is standing next to her sunflower. Tara is 1 metre tall.

Tick (✓) the height that you estimate the sunflower to be.

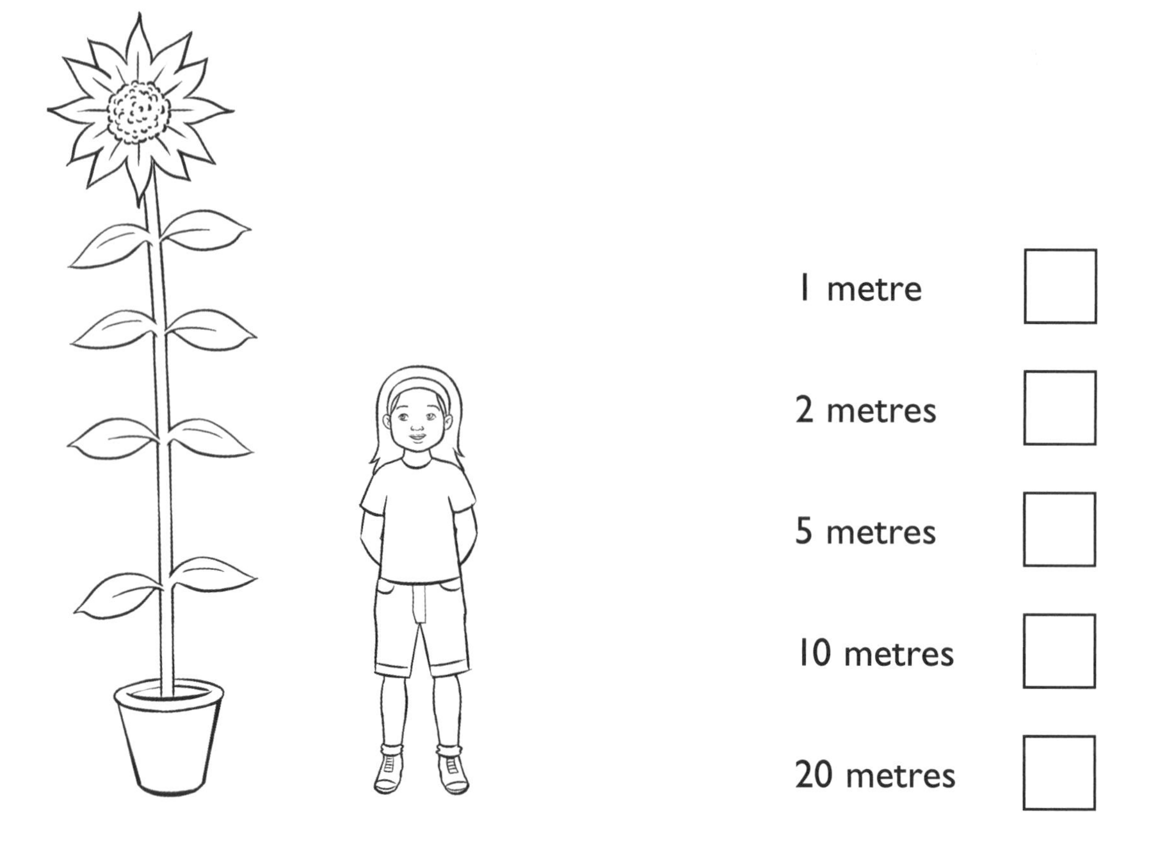

7 Look at the grid.

One of the squares is shaded.

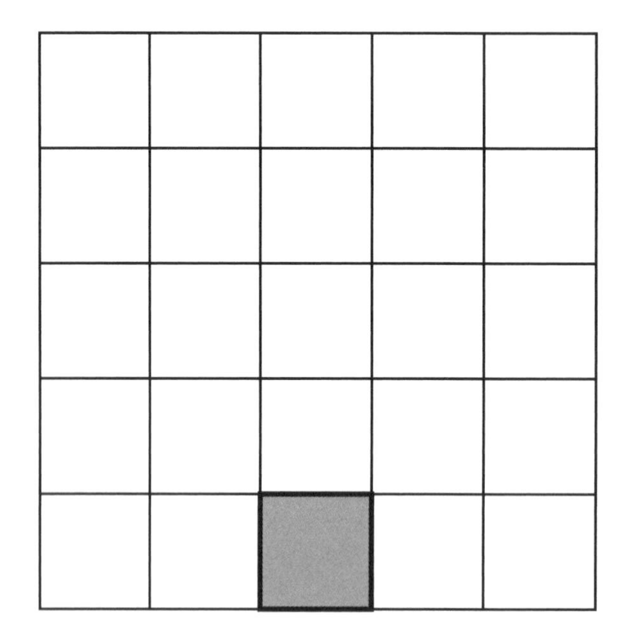

Start at the shaded square.

Move 3 squares up and 2 squares left.

Shade the square that you are now on.

8 Write the missing number.

An example has been done for you.

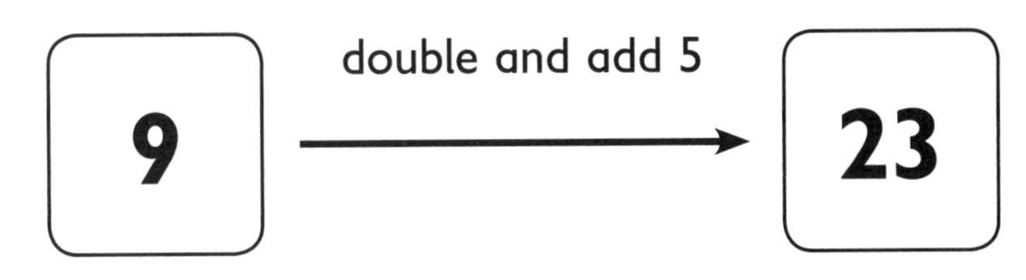

7 → double and add 5 → ☐

9

Yasmin has emptied her money box.

She needs 56p for a set of pens.

Write 2 different ways that Yasmin could pay for the pens using her coins.

1	2

2 marks

10 Write a number that is between these numbers.

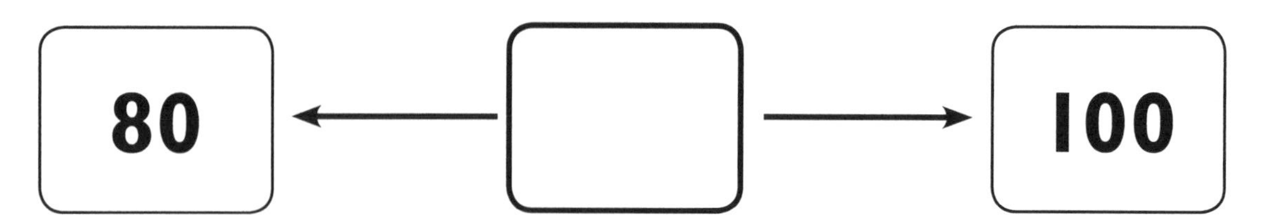

11 There are 50 balloons in a box.

17 balloons are green.

20 balloons are blue.

The rest of the balloons are yellow.

How many balloons are yellow?

Show your working

balloons

2 marks

12 Sort the numbers into this Venn diagram. Some of the numbers cannot be used.

35 16 40 15 27 36 53

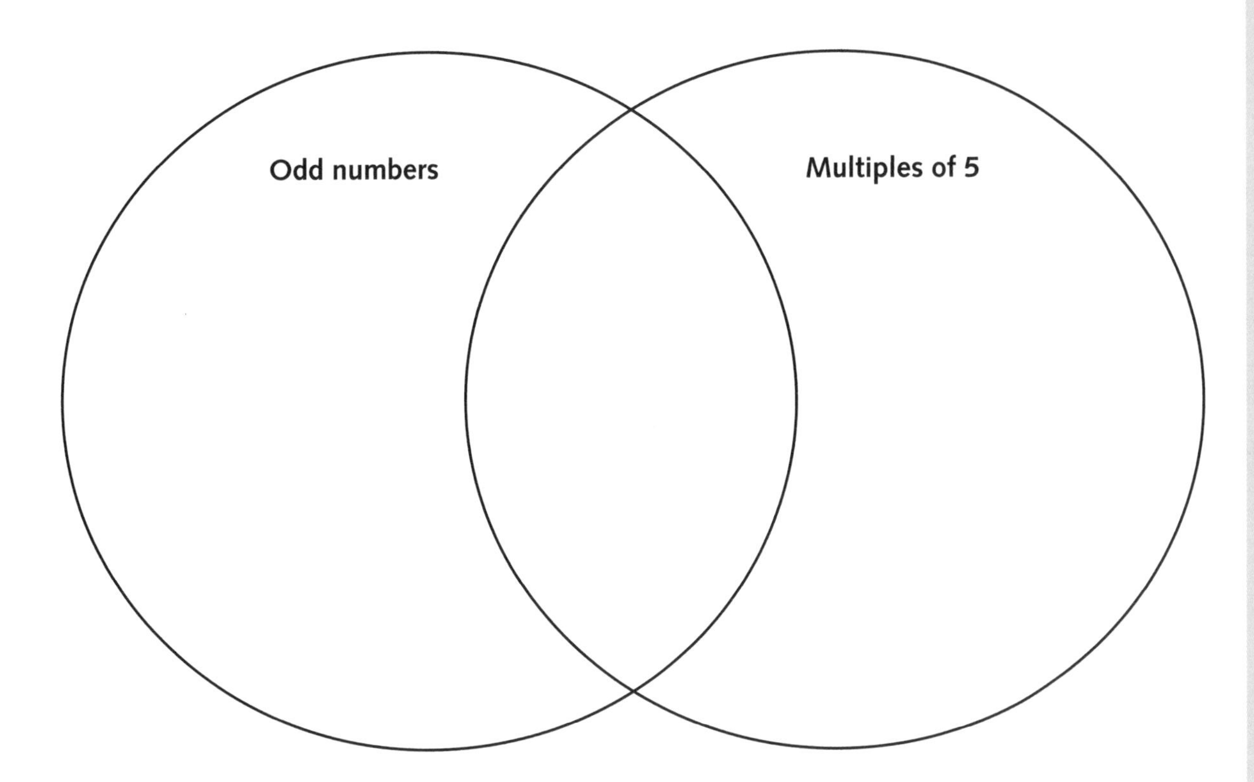

2 marks

13 Look at this number sequence.

1st	2nd	3rd	4th
18	21	24	27

What would the next number in this sequence be?

14 Ali went to the market to buy fruit.

Apples 15p

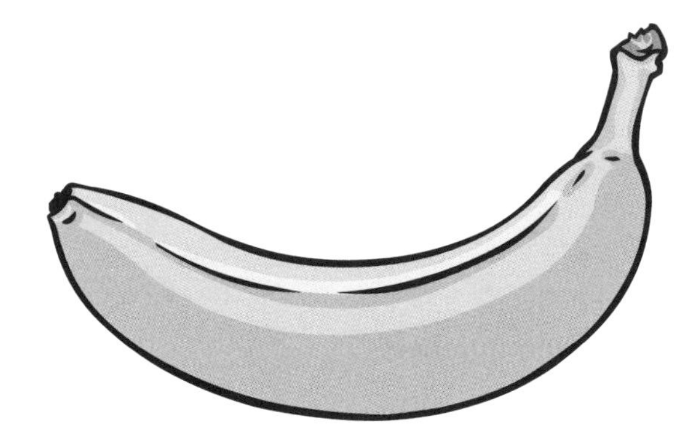

Bananas 25p

He wants to buy 2 bananas and 1 apple.

How much money will Ali need?

p

15 **Two** of these sentences describe true properties of a cylinder.

Tick (✓) the correct sentences.

Cylinders have 2 circular faces.

Cylinders have more than 4 corners.

Cylinders do not have right-angle vertices.

Cylinders have 3 triangular faces.

16 Fran made a tally of the birds visiting her garden.

Bird	Tally	Total
Robin	||||	
Finch	𝍸 𝍸 |	
Pigeon	𝍸	
Starling	𝍸 𝍸 ||	
Thrush	𝍸 |||	

Help Fran by adding the totals to her chart.

17 Look at this grid.

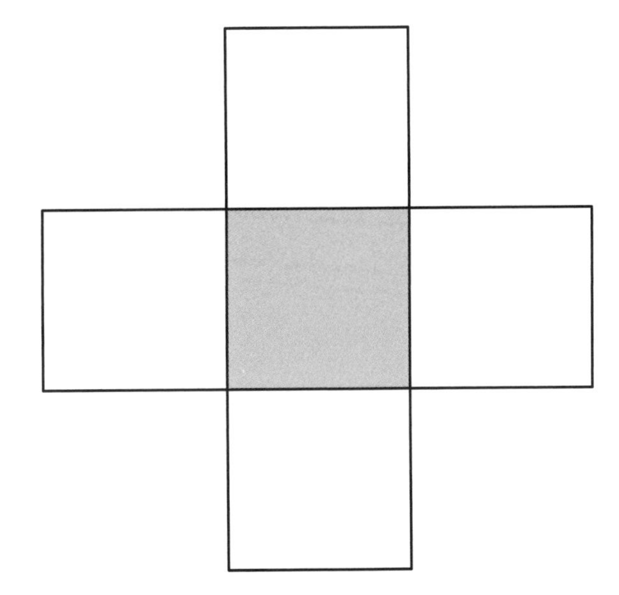

Put an X in the square above the shaded square.

Put a Y in the square below the shaded square.

18 Write an equivalent fraction.

One has been done for you.

$\frac{4}{4}$ = I whole

$\frac{3}{3}$ = ______

$\frac{2}{4}$ = ______

$\frac{3}{2}$ = ______

19 Eric wants to fill his fish tank.

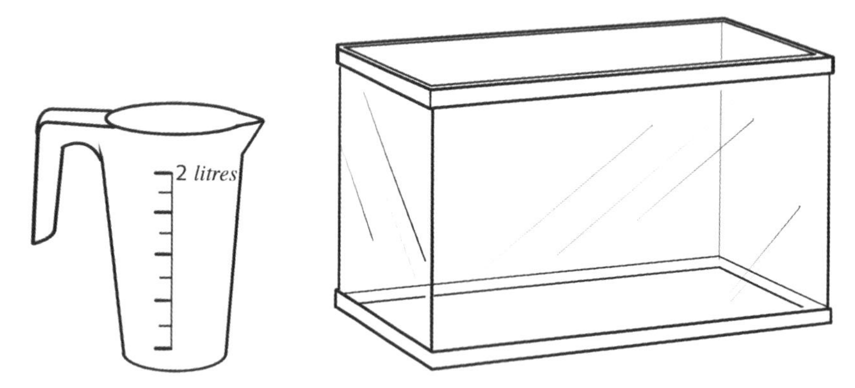

The fish tank holds 22 litres of water.

The jug holds 2 litres of water.

How many full jugs of water will Eric need?

20 Is each statement always, sometimes or never true?
Write **always**, **sometimes** or **never** in each box.

always **sometimes** **never**

Multiples of 3 end in 1.	
Multiples of 5 end in 2.	
Multiples of 10 end in 0.	

21 Draw the reflection of this shape using the dotted mirror line.

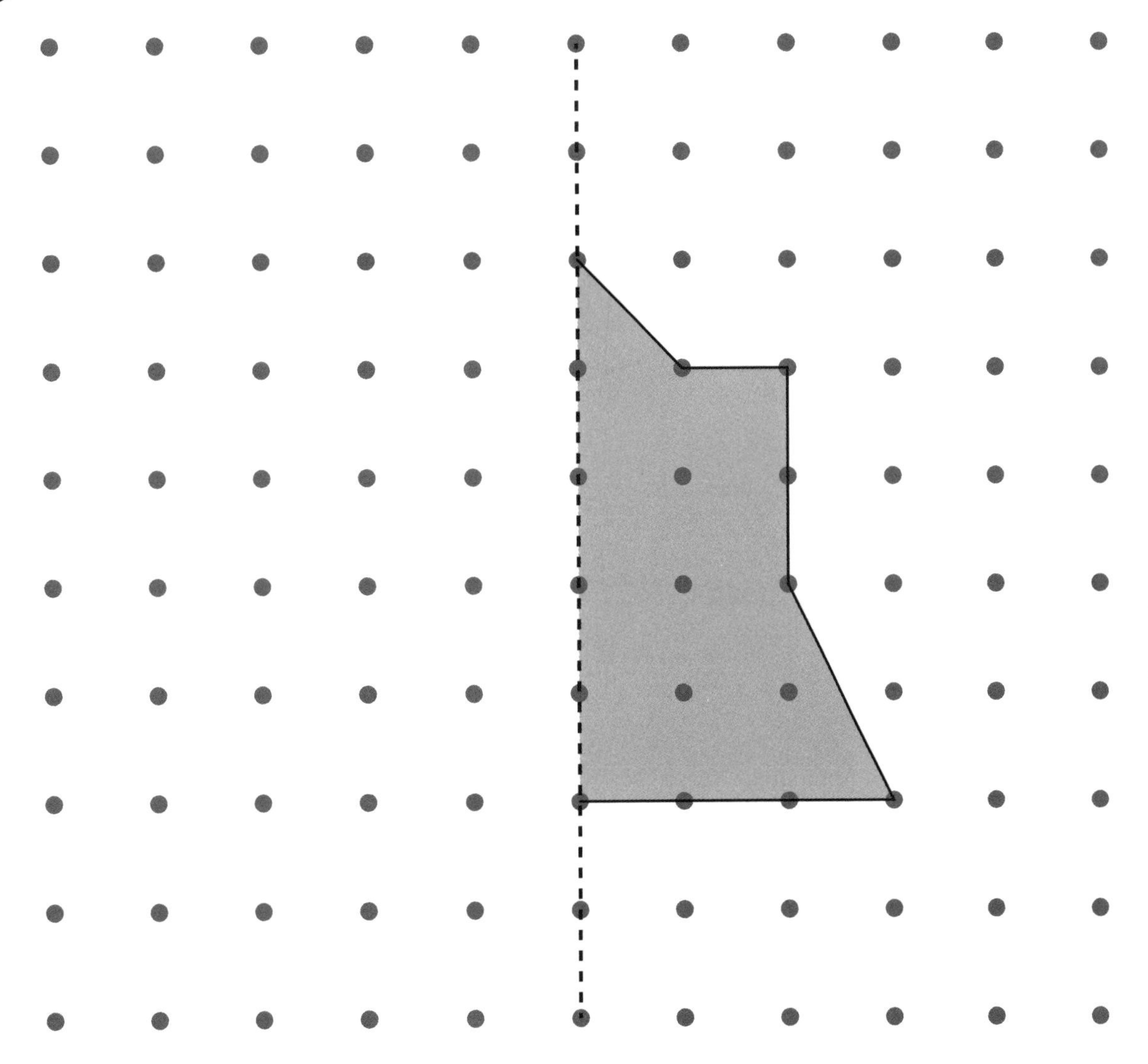

22 Add 10 to each of these numbers.

One has been done for you.

87	→	97
23	→	
47	→	
64	→	

23 Martina wants to make some biscuits.

For her biscuits Martina needs:

150 g flour

100 g butter

50 g dried fruit

50 g sugar

What is the total weight of Martina's ingredients?

[ ] g

24 Gabby walks to school.

She leaves home at half past seven.

Her journey takes 45 minutes.

a) Draw hands on the analogue (dial) clock to show the time that Gabby leaves home.

b) Draw hands on the analogue (dial) clock to show the time that Gabby arrives at school.

25 Liz's cat Smudge eats 2 packets of food each day.

How many packets of food does Smudge eat in 1 week?

Show your working

packets

2 marks

26 Shema has found that 2 crayons are the same length as the top of her book.

Estimate the number of crayons that Shema would need to go around the whole book.

crayons

27 Frankie buys a burger.

He pays with a £2 coin.

Frankie gets 50p change.

How much did the burger cost?

28 Draw the next three shapes in this sequence.

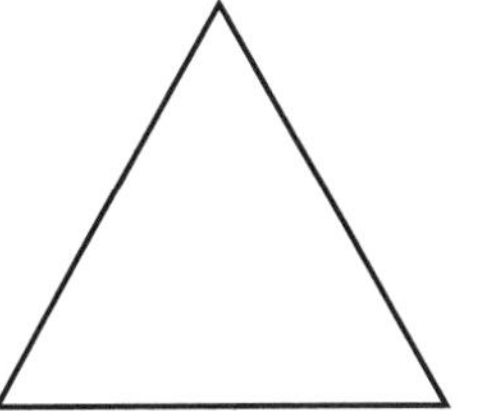

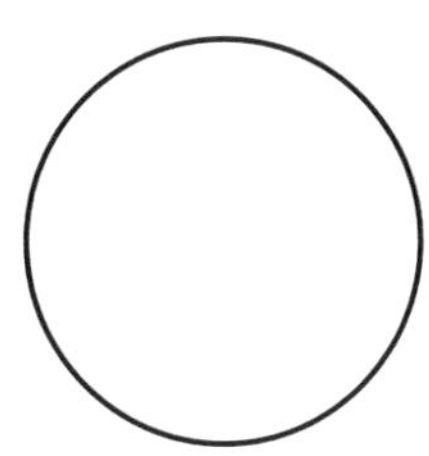
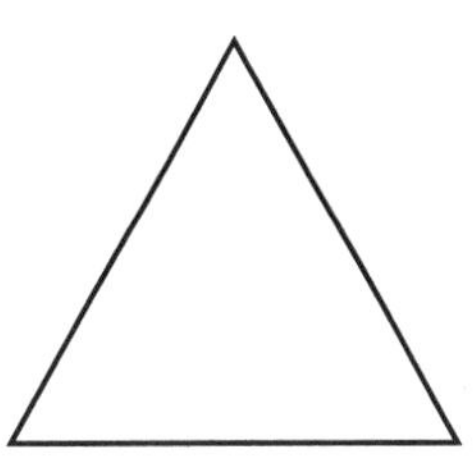
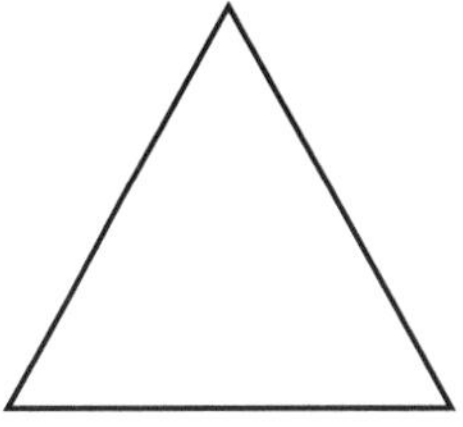

29 Use a ruler to draw a line that is 10 cm in length.

30 Write the answers in the boxes.

81 —take away 10→ ☐

46 —take away 10→ ☐

97 —take away 10→ ☐

Key Stage 1

SET C

Maths

PAPER 1

Maths

Paper 1: arithmetic

Time:

You have approximately **20 minutes** to complete this test paper.

Maximum mark	Actual mark
25	

First name	
Last name	

Practice question

$2 \times 3 = \square$

1

$6 \div 2 =$ ☐

2

☐ $= \frac{1}{2}$ of 24

3

$20 - 14 =$ ☐

4

$\square + 12 = 38$

5

$5 \times \square = 30$

6

$12 \div 2 = \square$

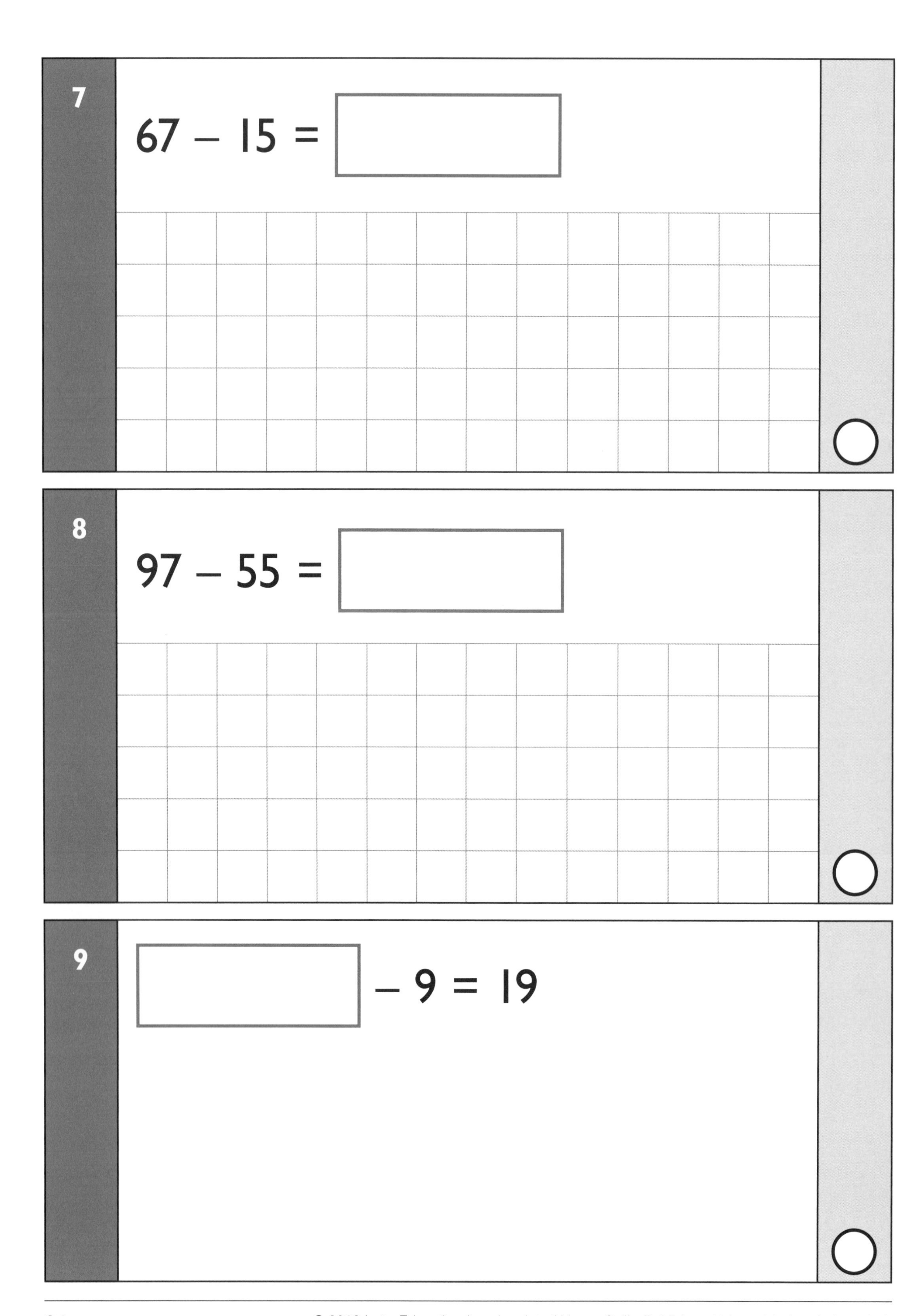
7
67 − 15 =
8
97 − 55 =
9
− 9 = 19

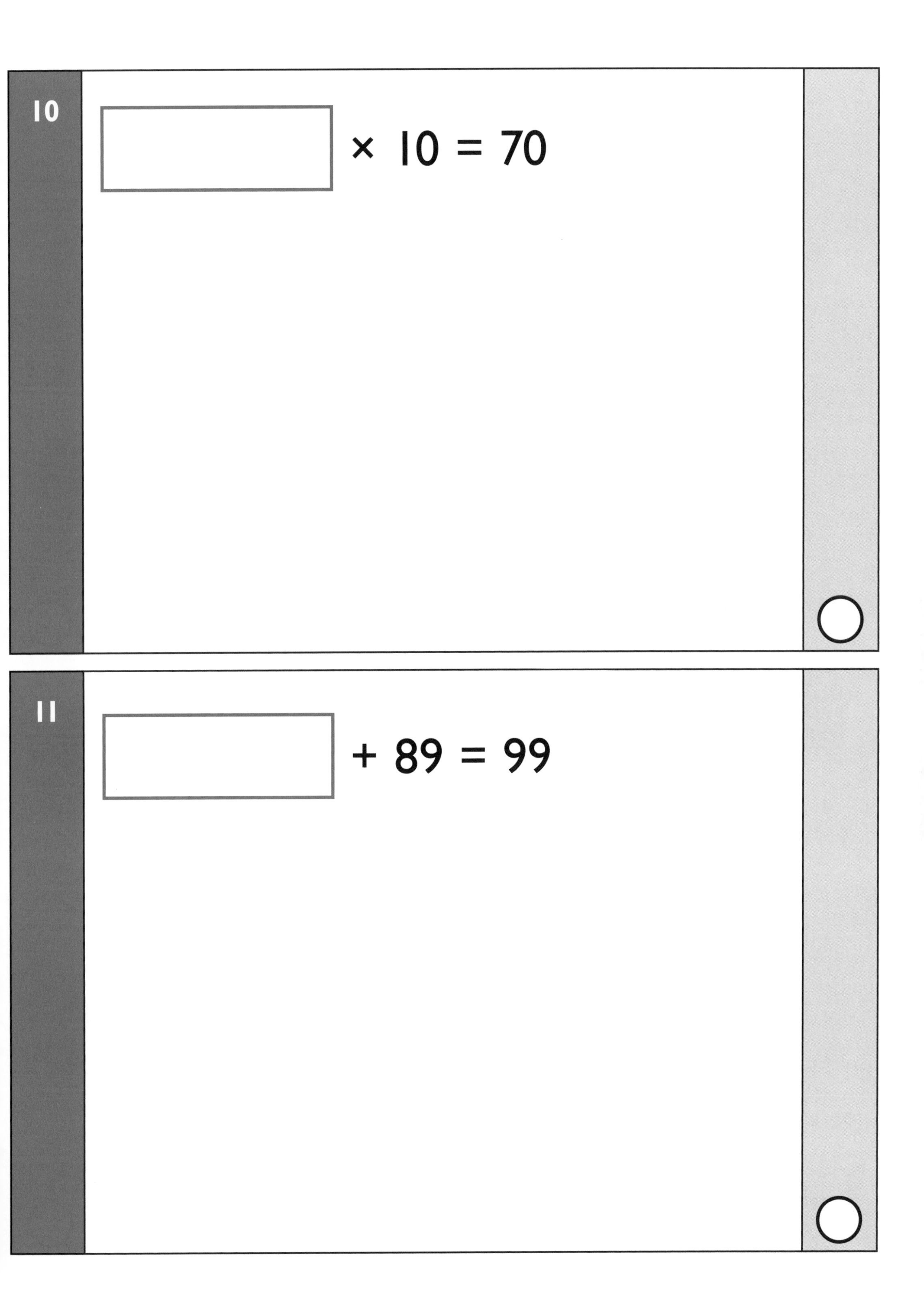
10
× 10 = 70
11
+ 89 = 99

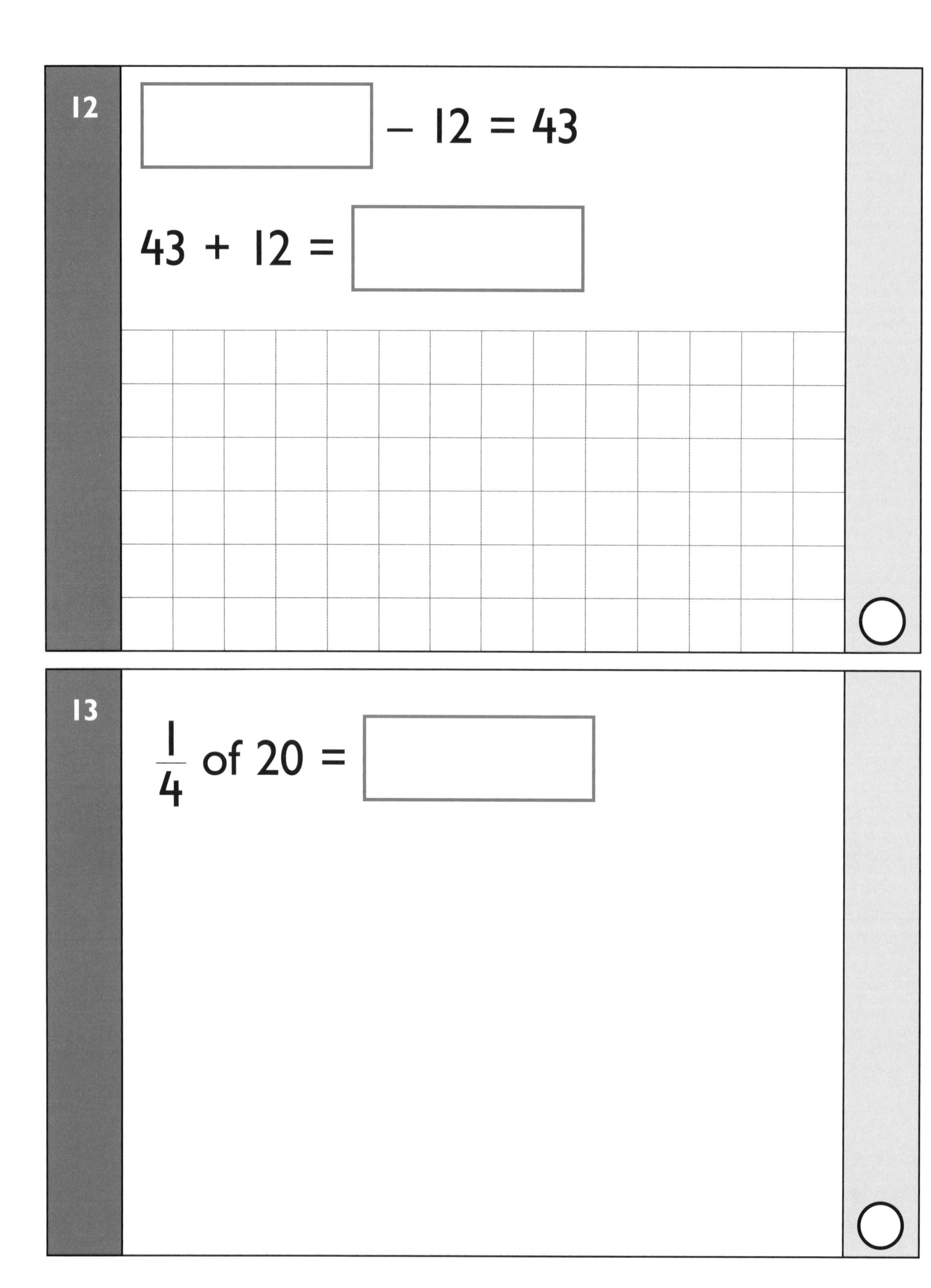
12
– 12 = 43
43 + 12 =
13
1/4 of 20 =

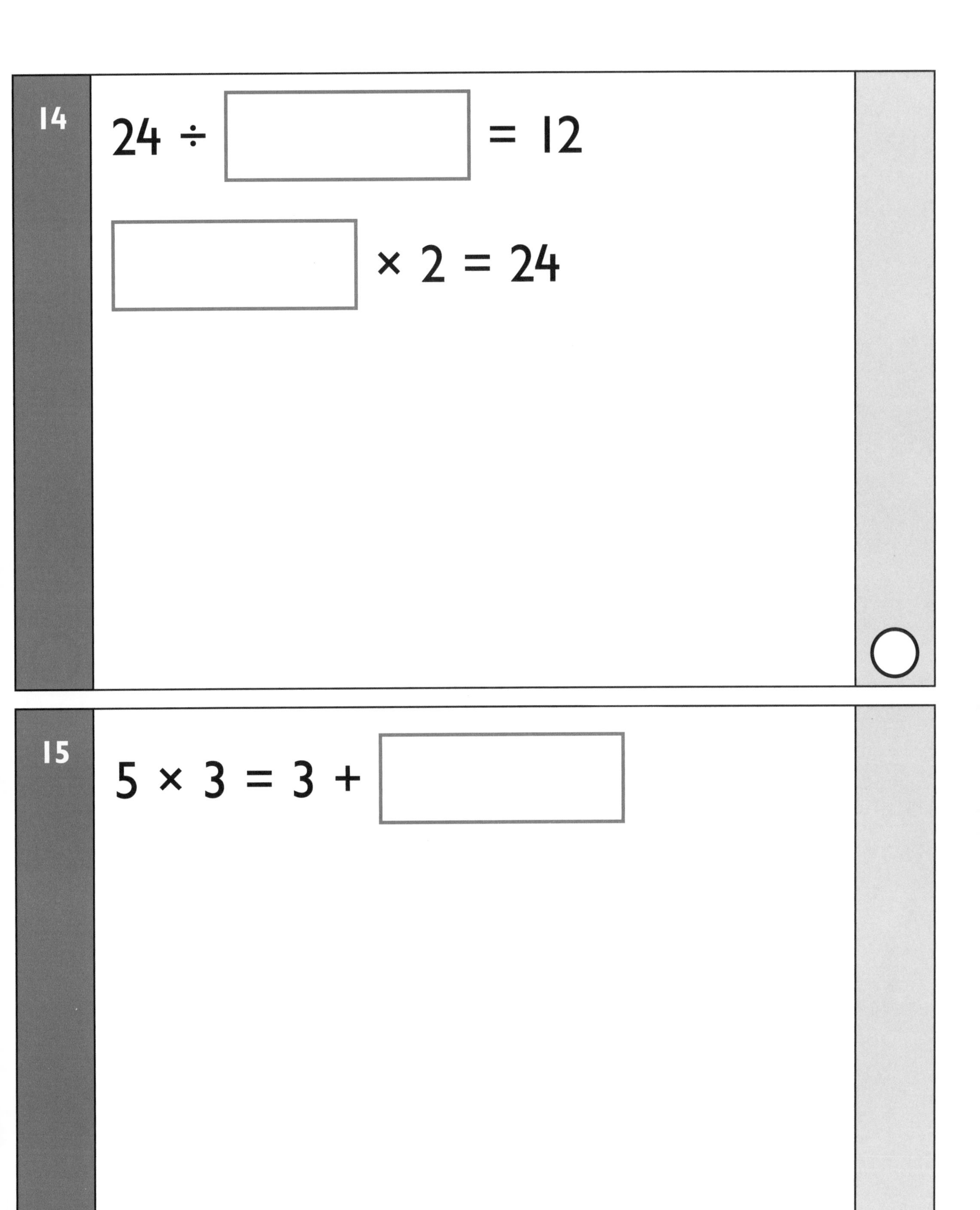
14
24 ÷ [ ] = 12
[ ] × 2 = 24
15
5 × 3 = 3 + [ ]

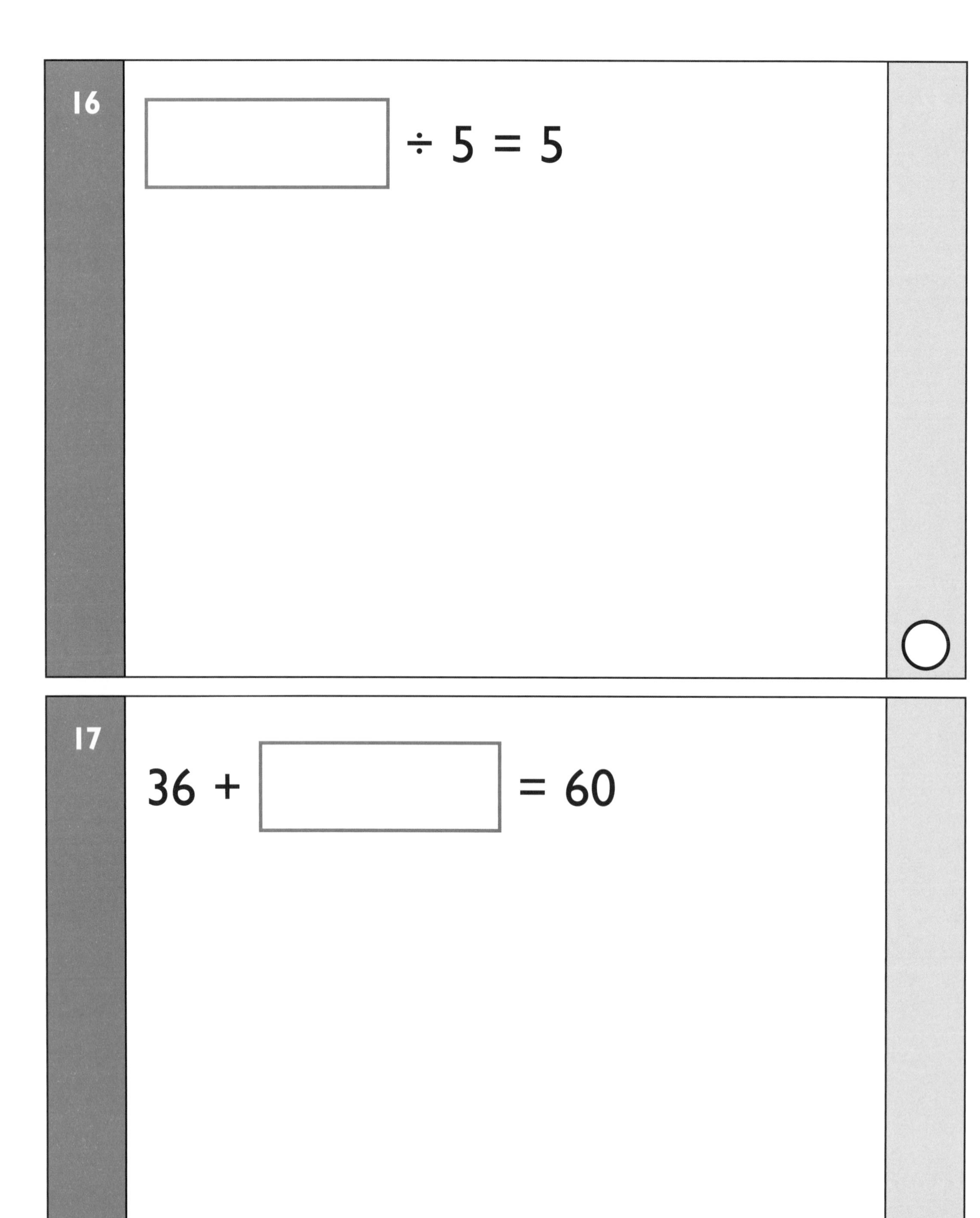
16
÷ 5 = 5
17
36 +
= 60

18

$\frac{1}{3}$ of 18 =

19

33 + 28 =

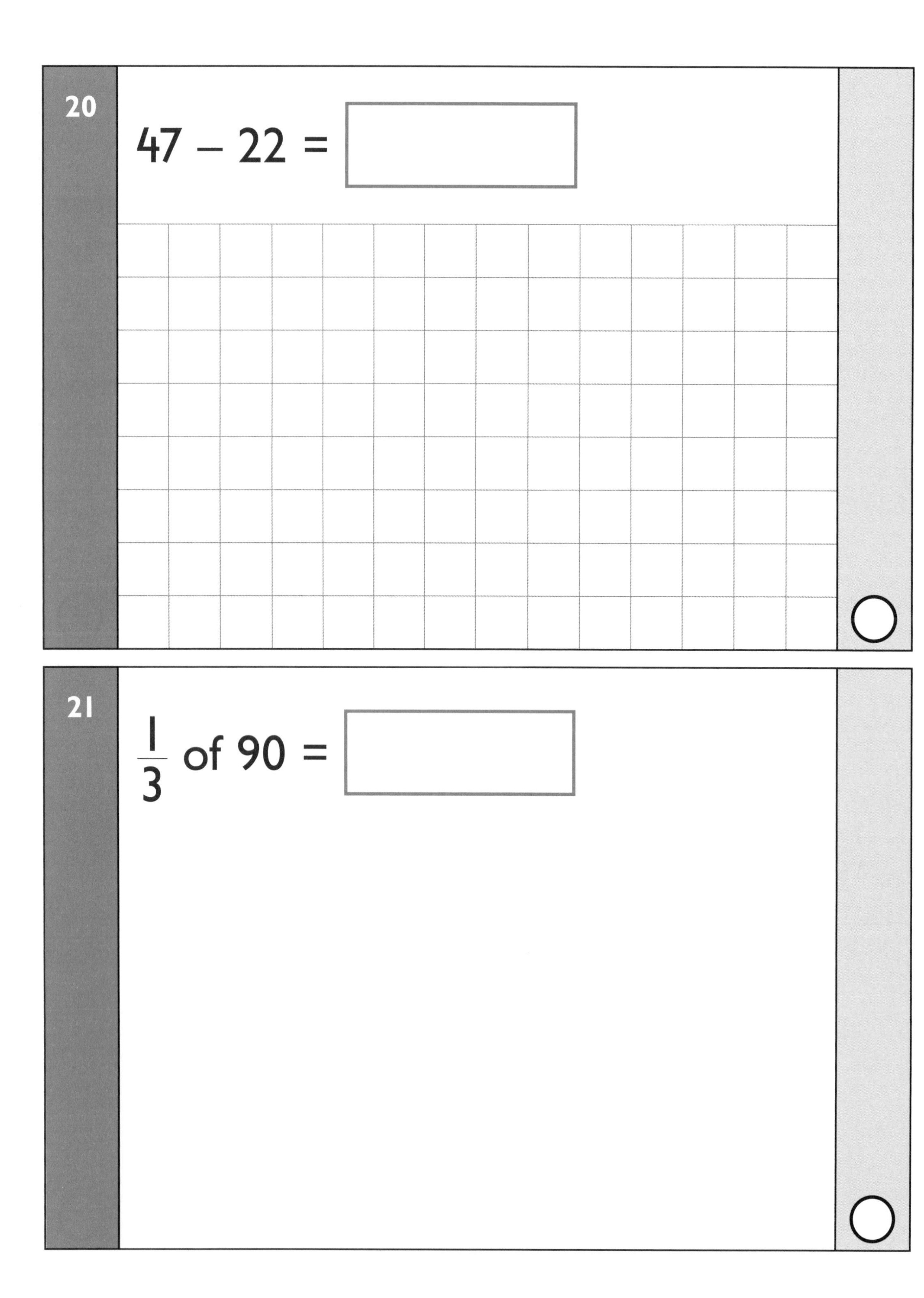
20
47 – 22 =
21
1/3 of 90 =

22

$20 \div \square = 4$

$\square \times 4 = 20$

23

$73 - 42 = \square$

24

$\frac{1}{3}$ of 99 = ☐

25

20 × ☐ = 100

Key Stage 1

SET
C

Maths

PAPER 2

Maths

Paper 2: reasoning

You will need to ask someone to read the instructions to you for the first five questions. These can be found on page 96. You can answer the remaining questions on your own.

Time:

You have approximately **35 minutes** to complete this test paper. This timing includes approximately 5 minutes for the aural questions.

Maximum mark	Actual mark
35	....

First name	
Last name	

Practice question

3	6	9	12	

1

p	p	p

2

3

children

4

vertices

5

litres **grams** **centimetres** **kilograms**

Now continue with the rest of the paper on your own.

6 Write the same number in each box to make this correct.

☐ + ☐ = 30

7 There are 10 satsumas in each bag and 8 more loose satsumas.

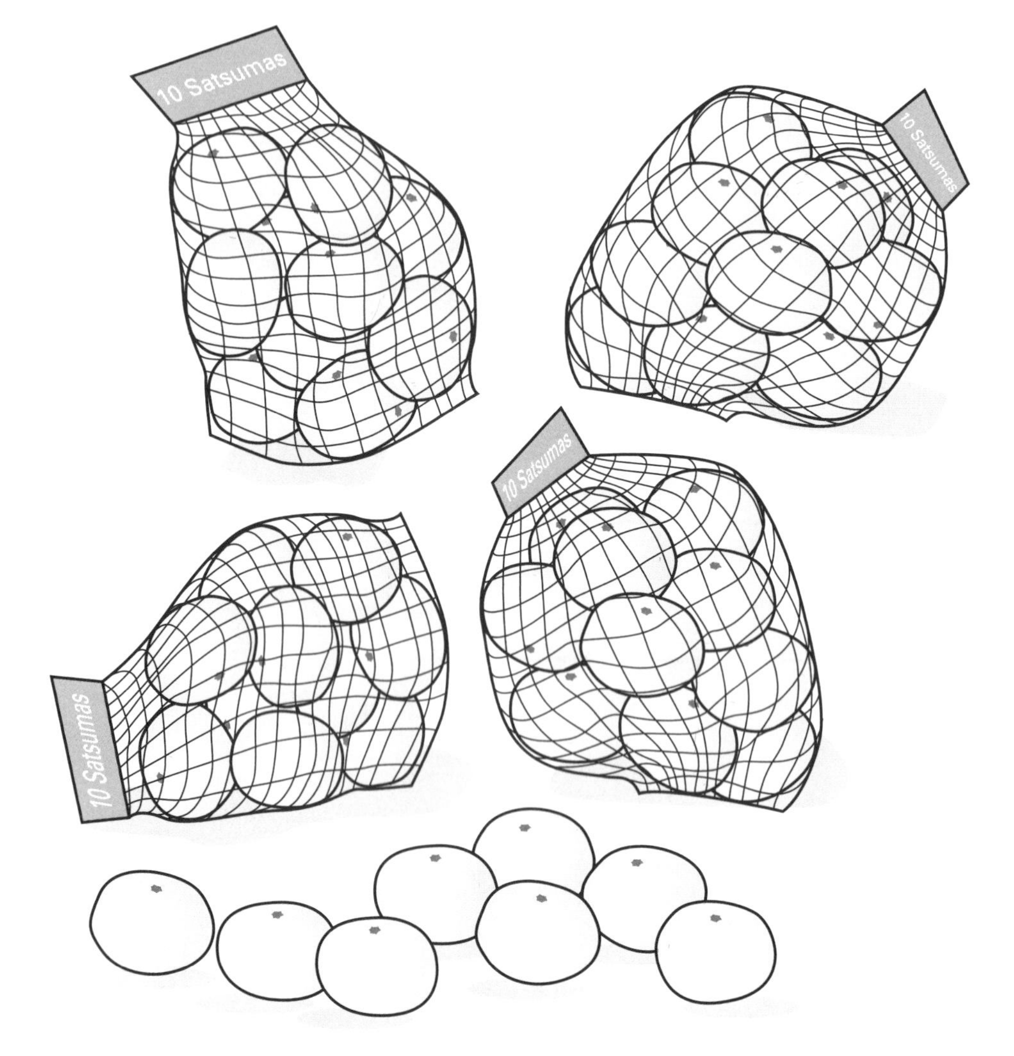

How many satsumas are there in total?

☐ satsumas

8 The two joined numbers add up to 18.

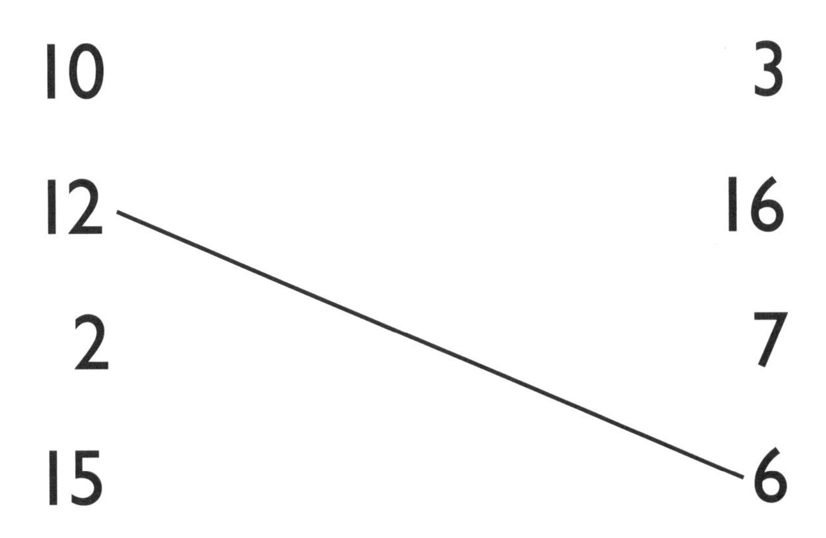

Join other numbers that have a total of 18.

9 Henry has weighed this bowl.

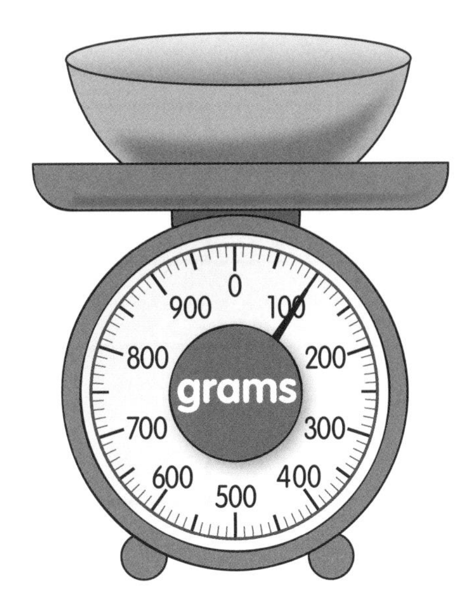

He has 5 bowls to take to his friends, so they can all have a share of ice cream.

How much do Henry's bowls weigh in total?

g

10 Ben worked out the correct answer to $50 \div 5$.

His answer was 10.

Show how Ben could have worked this out.

11 Fill in the missing numbers.

a)

15		21		27

b)

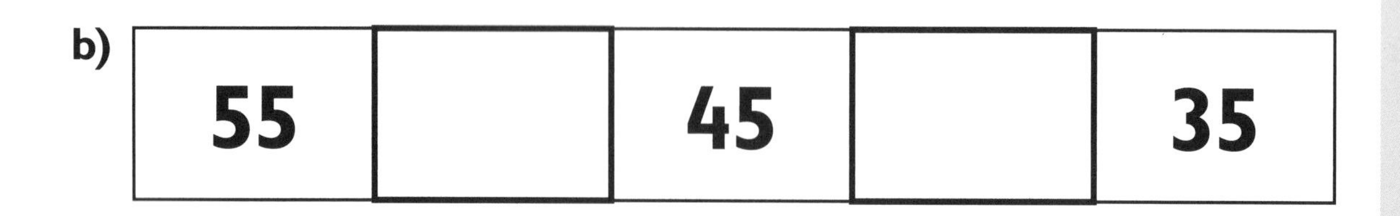

12 This shape has been divided into equal parts.

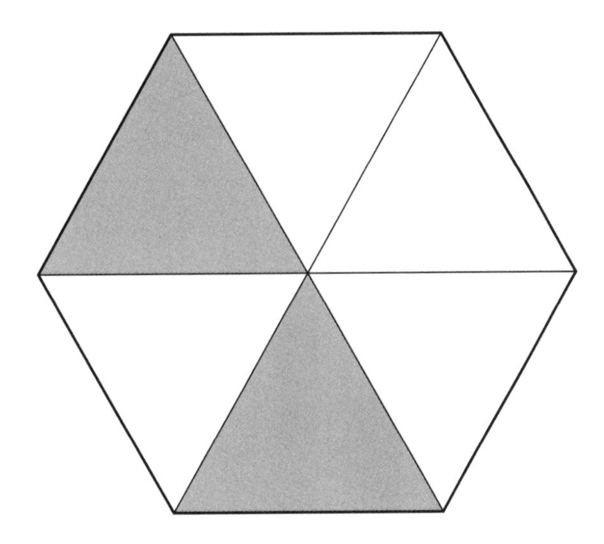

What fraction of the shape is shaded?

13 Circle the highest temperature.

20°C 22°C 18°C 21°C 12°C

14 Nina and her class have made a chart.

stands for 2 children

Hair colours

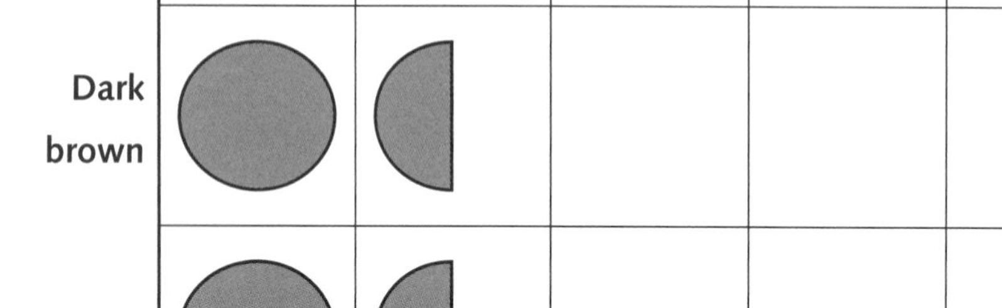

a) How many children have red hair?

children

b) Only one hair colour shows an **even** number of children.

Which colour is it?

15 Circle the coldest temperature.

1°C 3°C 5°C 0°C 10°C

16 Look at the number machine.

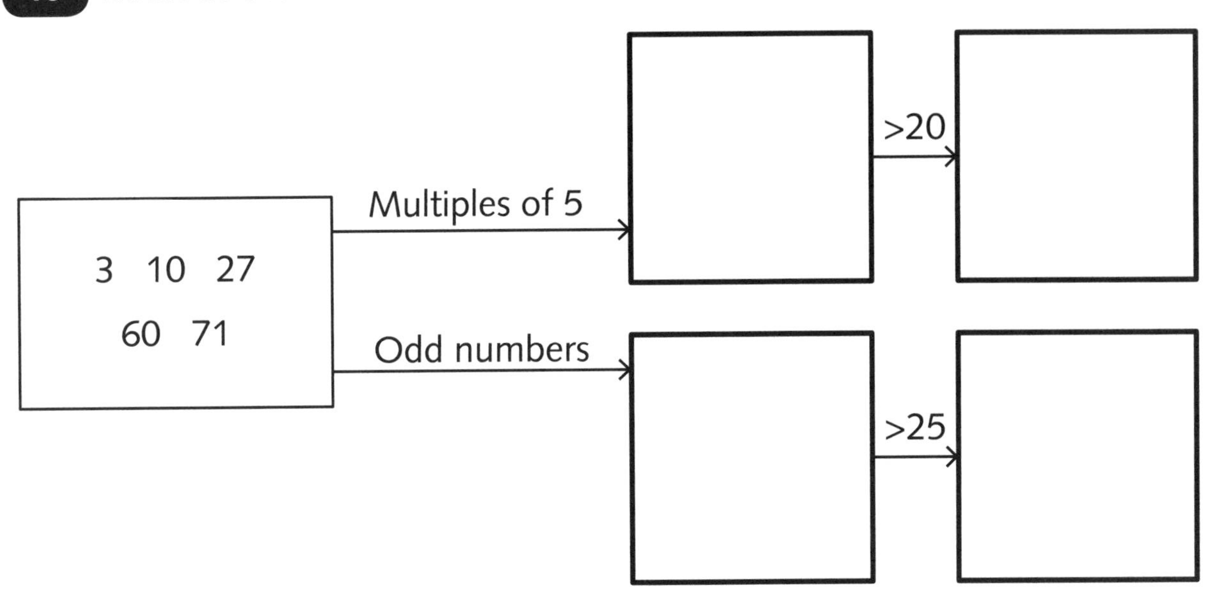

Put each number in the correct box(es).

2 marks

17 Tara has made a list of chores.

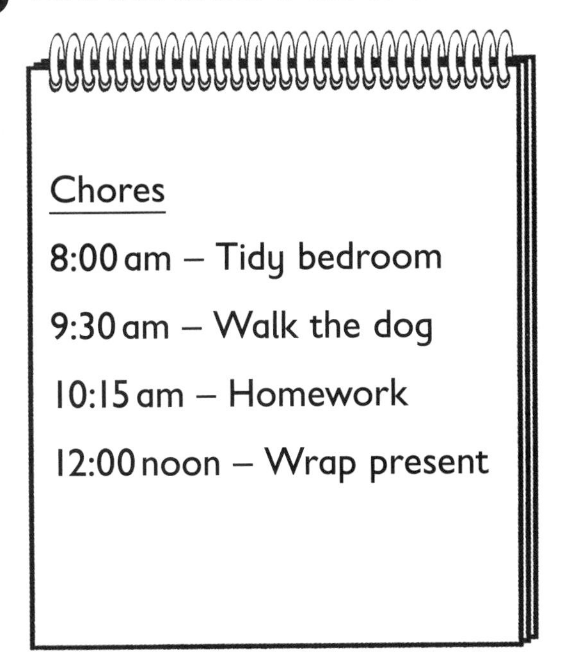

How long has Tara allowed to walk her dog?

18 Draw the missing lines.

One has been done for you.

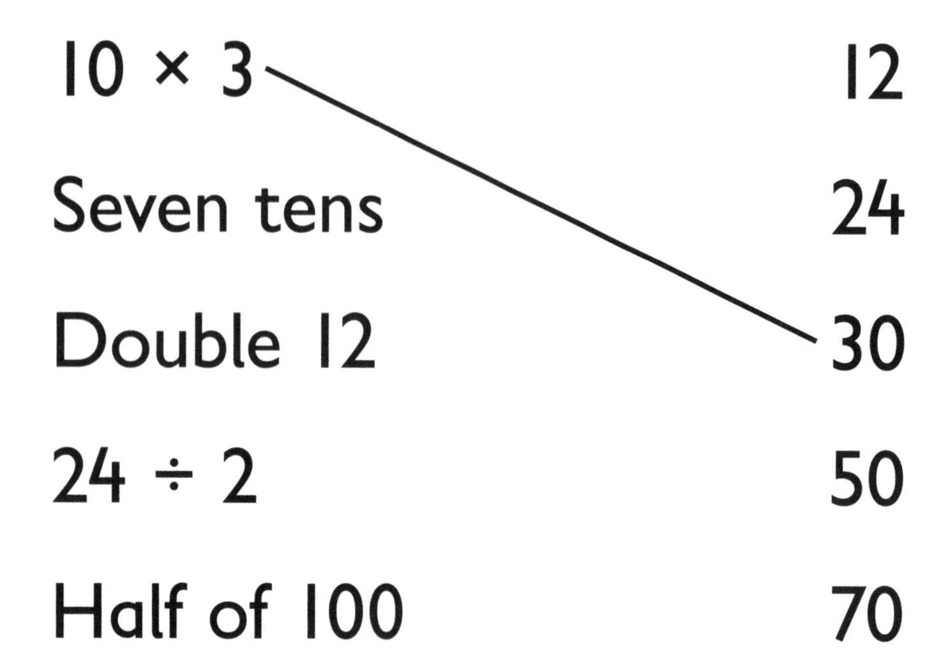

19 Alan is thinking of a number.

If he doubles his number and adds 7, the answer is 37.

Write the number that Alan is thinking of.

20 Shade the flash that is fourth from the right.

21 Hammed has found some card shapes.
He wants to make a cuboid.

Tick (✓) the 6 shapes that he should use.

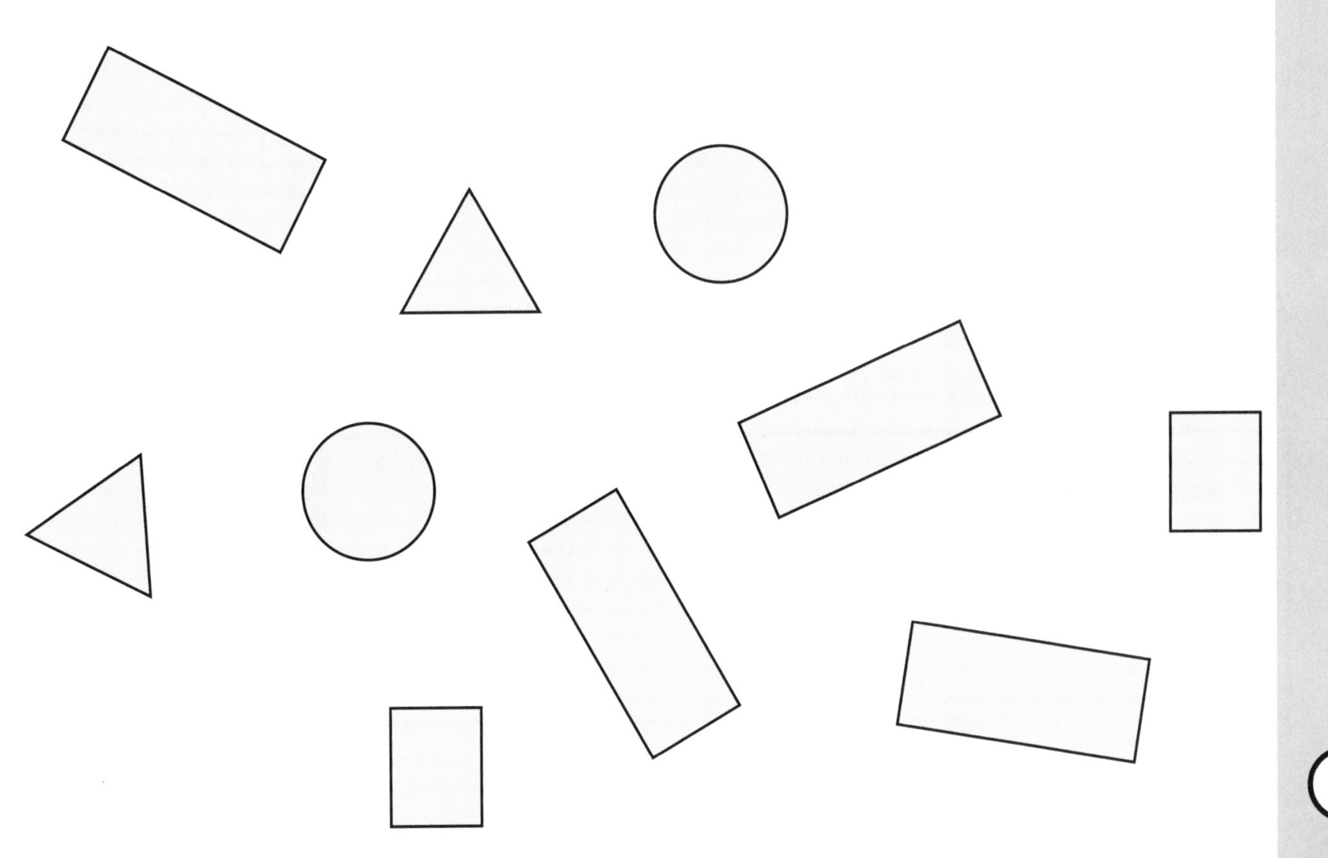

22 Alexi has emptied her money box.

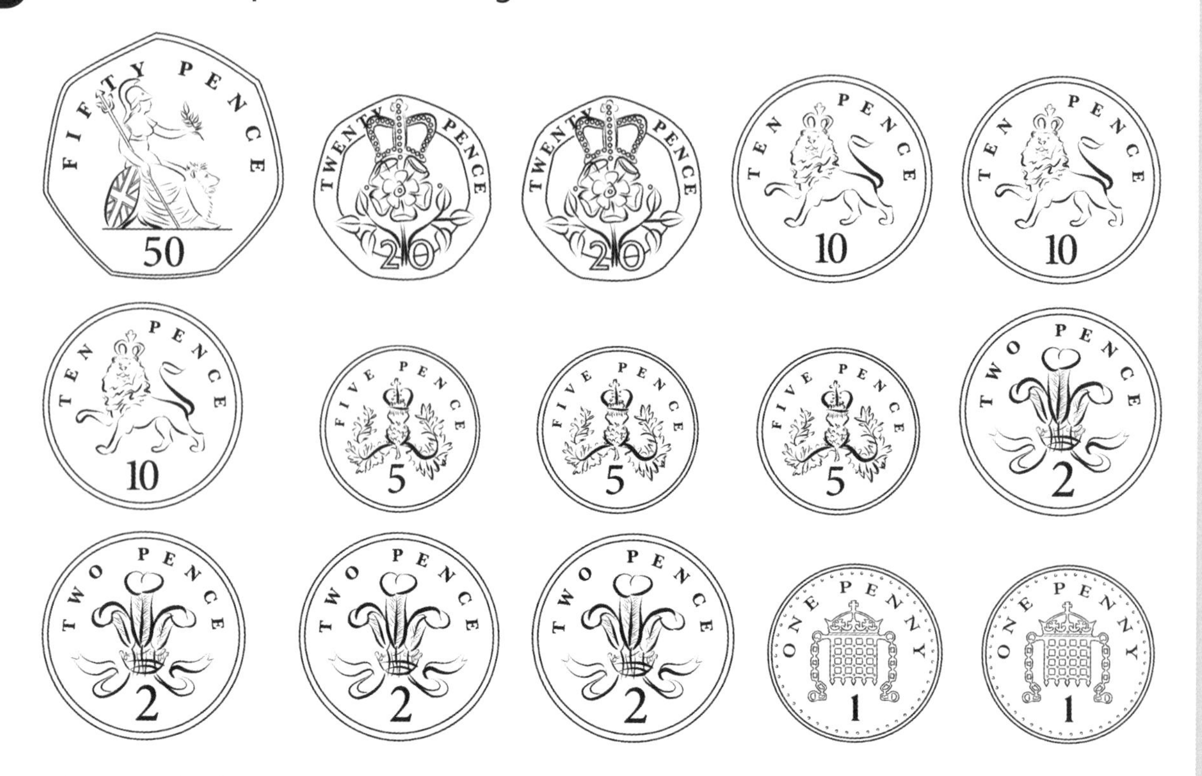

She needs **87p**.

Alexi takes the amount needed using the coins from her box.

Write which coins she could choose to find the exact amount needed.

23 What is the capacity of the bottle most likely to be?

Circle the correct amount.

5 litres **5 ml** **2 litres** **10 litres** **50 ml**

24 Look at the grid.

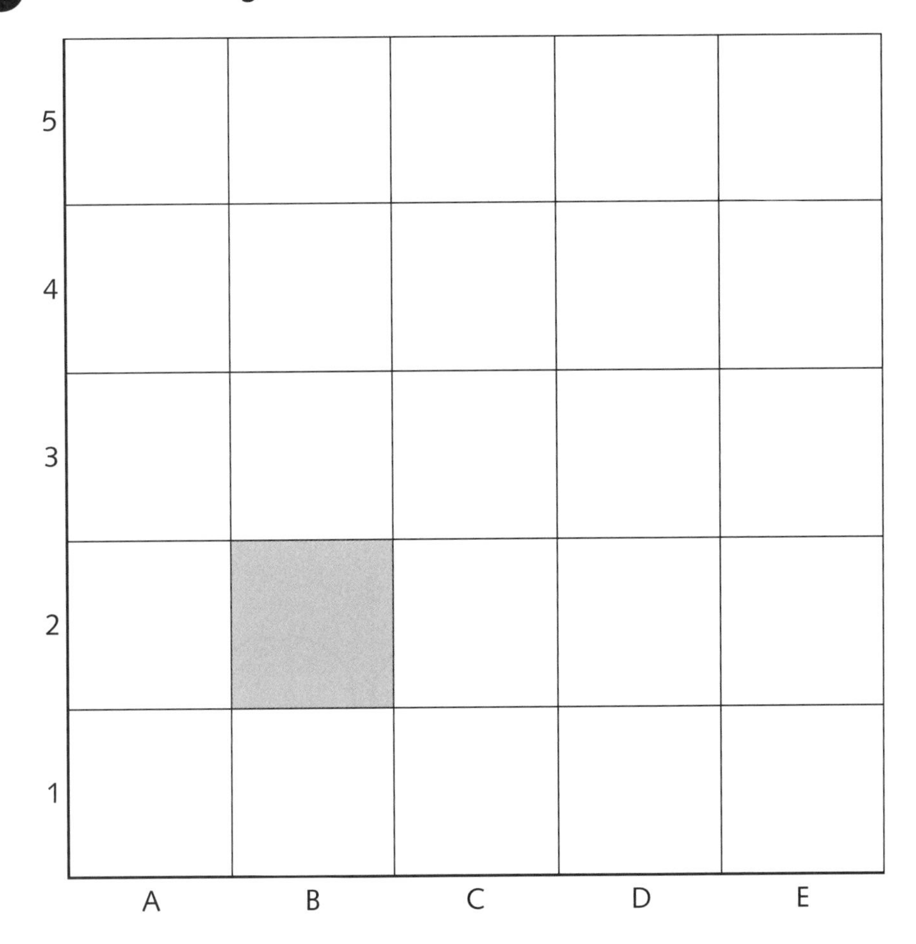

Square **B2** has been shaded.

Shade squares **C4**, **E3** and **D1**.

25

Jenny wants to buy a tablet computer that costs £90.

She saves £10 each week.

For how many weeks will she need to save until she has the right amount of money to buy the tablet computer?

weeks

26 Write the missing numbers.

a) $8 \times 10 = 94 - \square$

b) $50 \div 5 = 2 \times \square$

27 A group of children share 12 tomatoes.

Each child got 3 tomatoes.

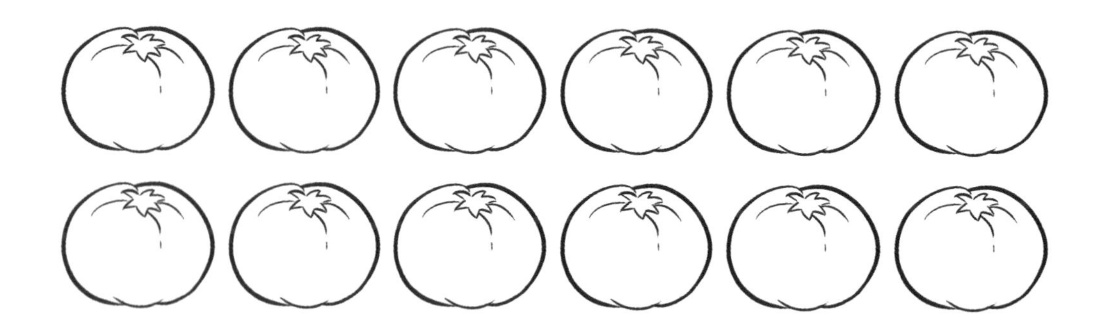

How many children shared the tomatoes?

28 Shade more squares so that $\frac{3}{4}$ of the whole is shaded.

29 Write the numbers in the correct positions.

410 **497** **458**

30 Class 2 have made a tally of the weather during March.

Weather in March

Weather	Tally	Total
Sunny	𝍸	
Showers	𝍸 |||	
Cloudy	𝍸 𝍸	
Rain	𝍸	
Snow	|||	

a) Count the tallies and write the totals in the correct column.

b) Add the missing data from the tally to the block graph.

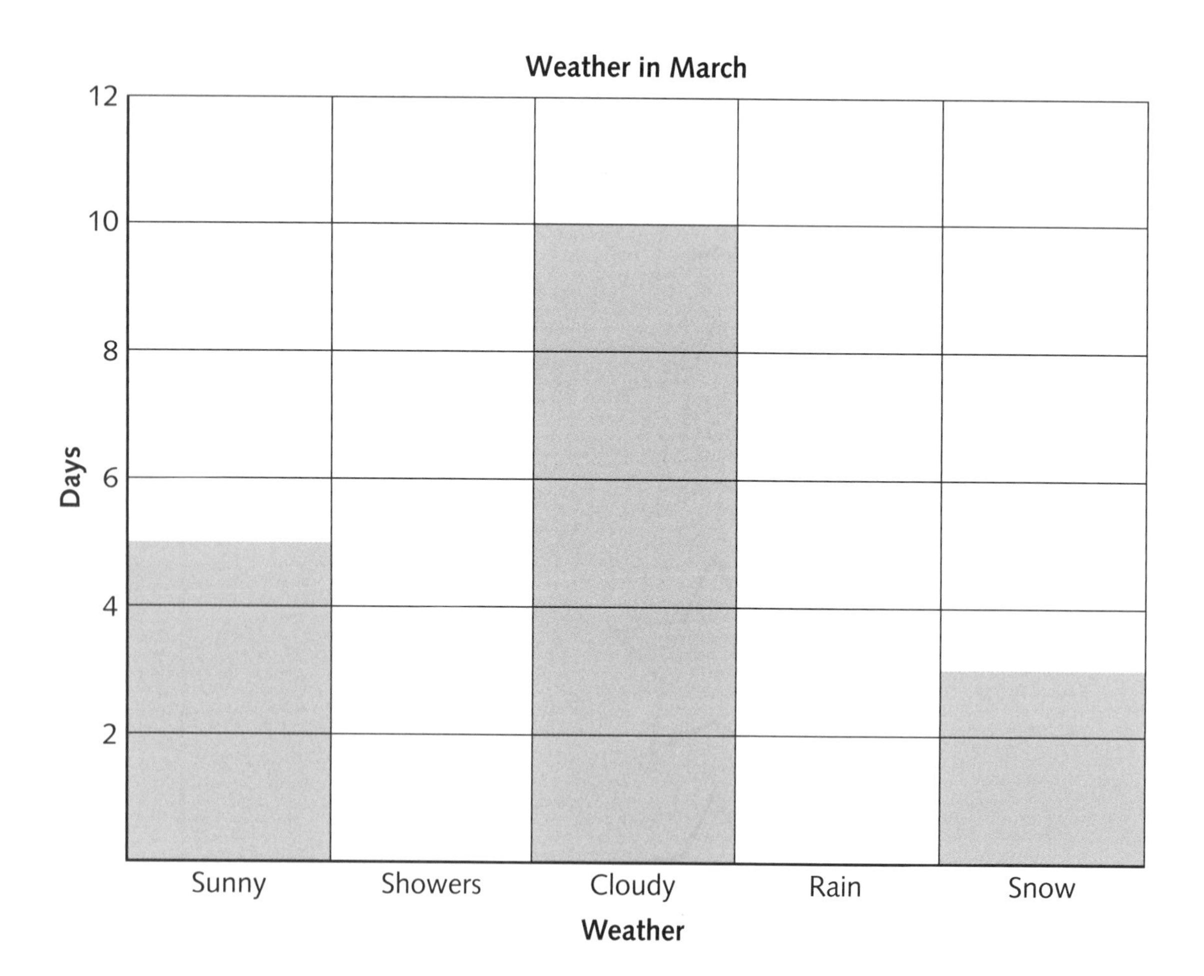

Answers and mark scheme

Notes to parents

Paper 1

These questions test calculating skills. An ability to work mentally will give a time advantage. Your child should be able to work with number facts to 20. (Children should know that 15 + 5 equals 20 and that 20 – 15 = 5; that 5 × 4 = 20 and that 20 ÷ 4 = 5.) Children should use this knowledge to extend their understanding of number facts to 100. Where the answer appears before the main content of the calculation, this tests your child's ability to use inverse (opposite) relationships between addition and subtraction. Knowing a fraction of a given quantity is an extension of your child's multiplication and division skills.

Paper 2

These questions test your child's ability to apply mathematics to problems and to reason, choosing an appropriate method to answer the question. The aural section will allow your child to become accustomed to the type of questions that they should expect and will experience during the actual tests at school.

The written questions test a wide variety of skills and the ability to choose a suitable mathematical strategy to answer the problems. Some of the questions provide an extra mark for showing how the problem was solved. This is an opportunity to see how your child is using the skills that are required to become a successful mathematician. Some of the questions require a range of skills, others are more straightforward.

Set A Paper 1

Practice question 4

1. 8 **(1 mark)**
2. 7 **(1 mark)**
3. 5 **(1 mark)**
4. 7 **(1 mark)**
5. 21 **(1 mark)**
6. 5 **(1 mark)**
7. 2 **(1 mark)**
8. 32 **(1 mark)**
9. 15 **(1 mark)**
10. 15 **(1 mark)**
11. 40 **(1 mark)**
12. 22 **(1 mark)**
13. 8 **(1 mark)**
14. 44 **(1 mark)**
15. 25 **(1 mark)**
16. 29 **(1 mark)**
17. 40 **(1 mark)**
18. 5 **(1 mark)**
19. 36 **(1 mark)**
20. 3 **(1 mark)**
21. 40 **(1 mark)**
22. 7 **(1 mark)**
23. 60 **(1 mark)**
24. 5 **(1 mark)**
25. 24 **(1 mark)**

Set A Paper 2

Practice question A group of 6 caterpillars circled.

1. 19 **(1 mark)**
2. 25 stickers **(1 mark)**
3. (3 kg) **(1 mark)**
4. 45p **(1 mark)**
5. **(1 mark)**

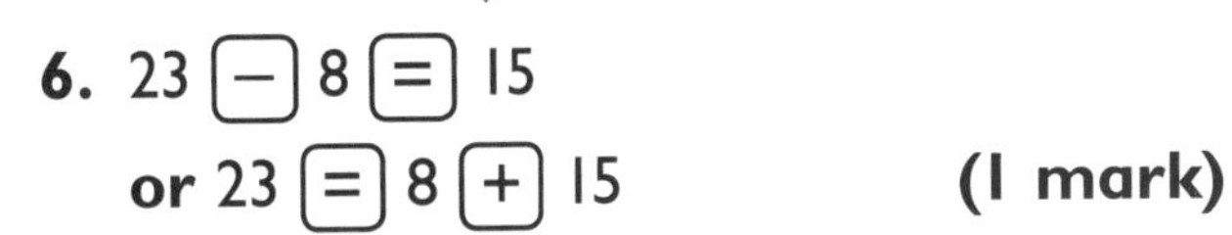

6. 23 [–] 8 [=] 15
or 23 [=] 8 [+] 15 **(1 mark)**

7. **45** is less than **57** ✓
34 is more than **27** ✓
28 is less than **34** ✓
59 is more than **7** ✓ **(1 mark)**

8. 18 children **(1 mark)**

9. (16) (76) (50) **(1 mark)**

10.

2 × 5	10
4 × 5	20
6 × 5	**30**

(1 mark)

11.

14	16	29	37	78	98	100

least most

(1 mark)

12. 15 bananas circled.

This could be 3 lots of 5, 15 as a group or any 15 bananas in any combination. **(1 mark)**

13. 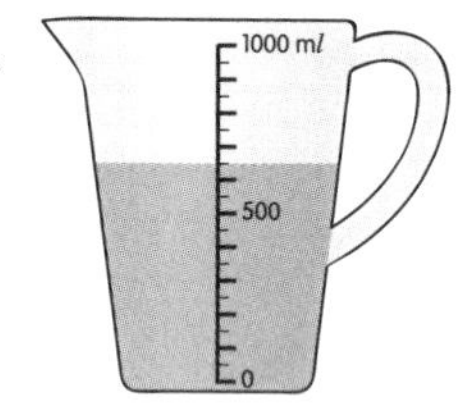

(1 mark)

14. 15 10p coins **(1 mark)**

15.

Numbers less than 30	Numbers more than 30
16 28	51 43 32 34

(1 mark)

16. *Either 1 of the following multiplications and either 1 of the following divisions:*

2 × 4 = 8 4 × 2 = 8

8 ÷ 4 = 2 8 ÷ 2 = 4 **(1 mark)**

17.  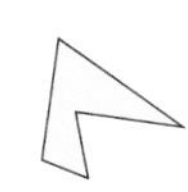 **(1 mark)**

 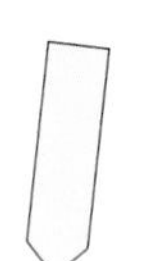

18. 20p

Any method that shows the amount:

2 × 50p = £1 3 × 10p = 30p

£1 + 30p = £1.30

or

50p + 50p + 10p + 10p + 10p = £1.30

£1.50 – £1.30 = 20p

(2 marks for correct answer. Award 1 mark for using appropriate method but wrong answer given.)

19. 85 pieces **(1 mark)**

20. a) Snake column shows 3.

Note: It is important that the graph shows an understanding of each square representing 2 items and that the 3 snakes are shown as a solid lower block of 2 and a split upper block counting as 1.

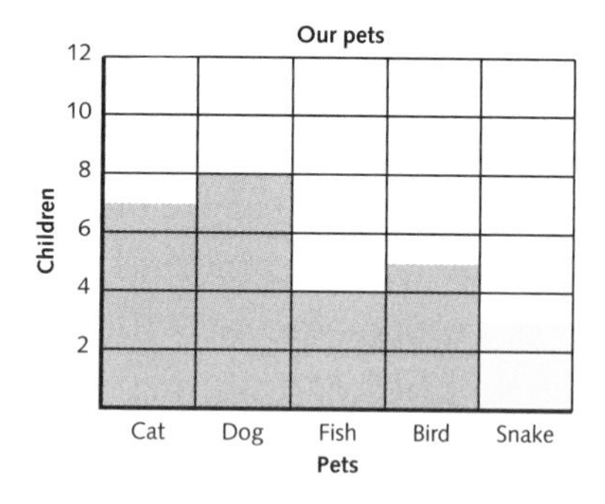

(1 mark)

b) 3 children **(1 mark)**

21. a)

2	5	**8**	11	**14**	17

(1 mark)

b)

24	**21**	**18**	15	**12**	9

(1 mark)

22. 16°C **(1 mark)**

23. a) 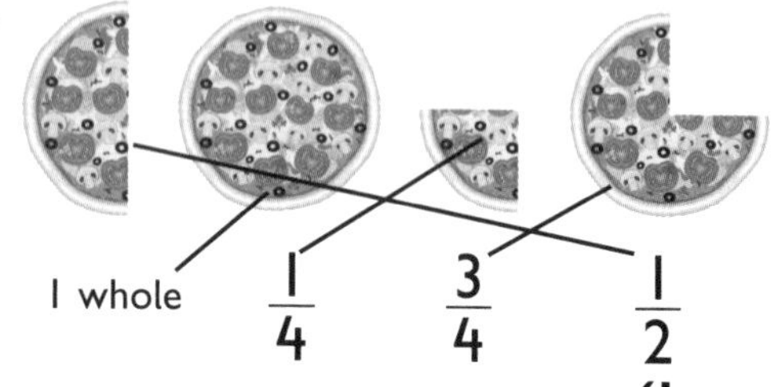

(1 mark)

b) 2 whole pizzas **(1 mark)**

24.   

(1 mark)

25. *Any number between* ***65*** *and* ***75*** *is acceptable.* **(1 mark)**

26.  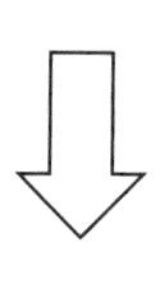  

(1 mark)

27. 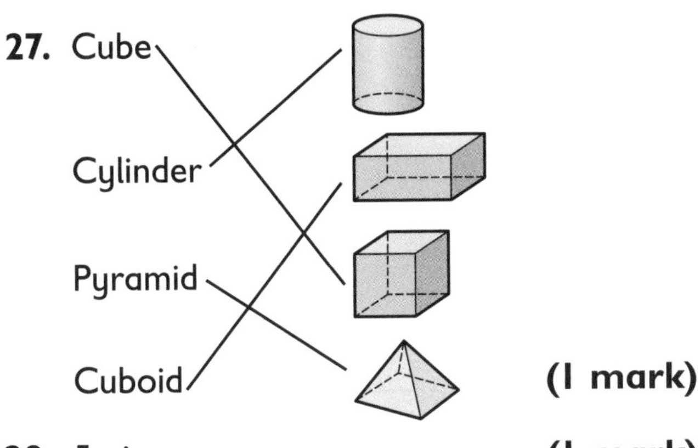

(1 mark)

28. 5 pieces **(1 mark)**

29. 6 packs **(1 mark)**

30. Vertices marked as shown:

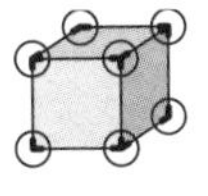

(1 mark)

31.

(1 mark)

Set B Paper 1

Practice question 12

1. 100 **(1 mark)**

2. 15 **(1 mark)**

3. 19 **(1 mark)**

4. 10 **(1 mark)**

5. 14 **(1 mark)**

6. 9 **(1 mark)**

7. $40 + \boxed{60} = 100$ **(1 mark: both**
$100 = 60 + \boxed{40}$ **correct for 1 mark)**

8. 7 **(1 mark)**

9. 100 **(1 mark)**

10. 10 **(1 mark)**

11. 25 **(1 mark)**

12. 31 **(1 mark)**

13. 20 **(1 mark)**

14. 10 **(1 mark)**

15. 5 **(1 mark)**

16. 0 **(1 mark)**

17. 30 **(1 mark)**

18. 17 **(1 mark)**

19. 75 **(1 mark)**

20. 30 **(1 mark)**

21. 42 **(1 mark)**

22. 25 **(1 mark)**

23. 40 **(1 mark)**

24. 22 **(1 mark)**

25. 2 **(1 mark)**

Set B Paper 2

Practice question 50 stickers

1. Accept 307 *or* three hundred and seven **(1 mark)**

2. May **(1 mark)**

3. 13 **(1 mark)**

4. 6 faces **(1 mark)**

5. (2 hours) **(1 mark)**

6. 2 metres ✓ **(1 mark)**

7. 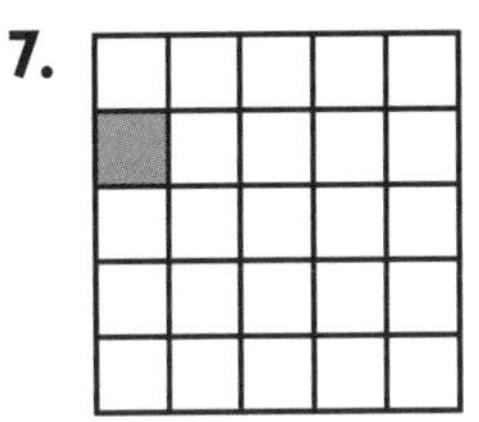

(1 mark)

8. 19 **(1 mark)**

9. *Any 2* ***different*** *combinations of available coins that equal 56p, e.g.* 50p + 5p + 1p

10p + 10p + 20p + 5p + 5p + 2p + 2p + 2p

or other variants.

Do not *accept multiples of coins that are* ***not*** *available, or amounts not represented by actual coins, e.g.*

50p + 6p (*A 6p coin does not exist*)

20p + 20p + 10p + 5p + 1p (*Only 1 × 20p is available*)

(2 marks: 1 mark for each correct combination)

10. *Any number from 81–99 inclusive.* **(1 mark)**

11. 13 balloons

Any viable method shown, e.g.

20 + 17 = 37 50 – 37 = 13

or

50 – 30 = 20 20 – 7 = 13

(2 marks for correct answer. Award 1 mark for using appropriate method but wrong answer given.)

12.

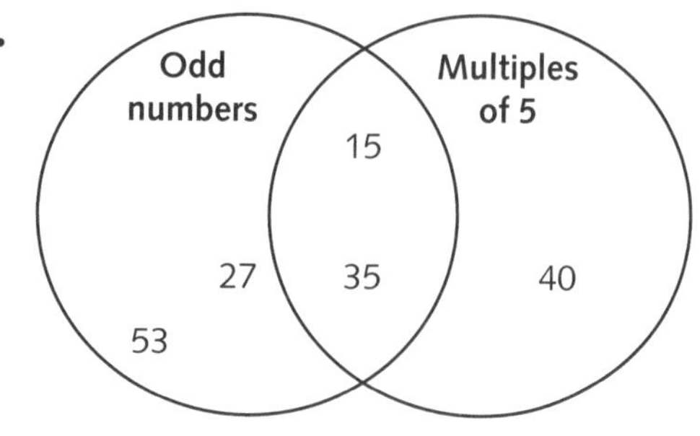

(2 marks: 1 mark for one of the circles correctly completed)

13. 30 **(1 mark)**

14. 65p **(1 mark)**

15. Cylinders have 2 circular faces. ✓

Cylinders do not have right-angle vertices. ✓ **(1 mark)**

16.

Bird	Tally	Total
Robin	||||	4
Finch	卌 卌 |	11
Pigeon	卌	5
Starling	卌 卌 ||	12
Thrush	卌 |||	8

(1 mark)

17.

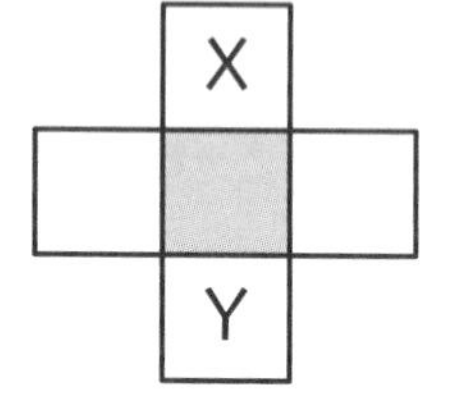

(1 mark)

18. *Any suitable answers, e.g.*

$\frac{3}{3}$ = 1 whole

$\frac{2}{4}$ = $\frac{1}{2}$ *or* half

$\frac{3}{2}$ = $1\frac{1}{2}$ *or* one-and-a-half **(1 mark)**

19. 11 jugs **(1 mark)**

20.

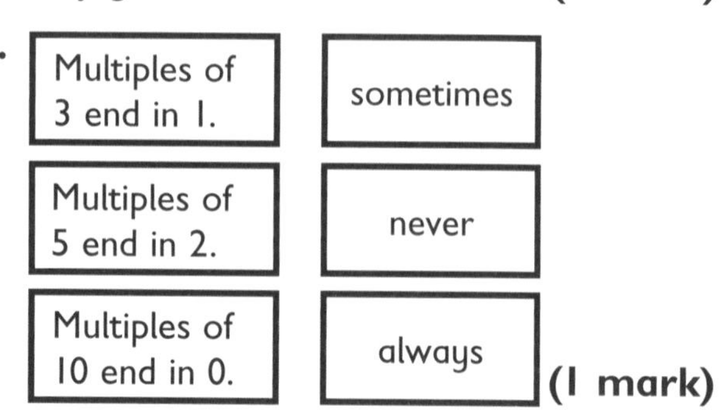

Multiples of 3 end in 1. — sometimes

Multiples of 5 end in 2. — never

Multiples of 10 end in 0. — always

(1 mark)

21.

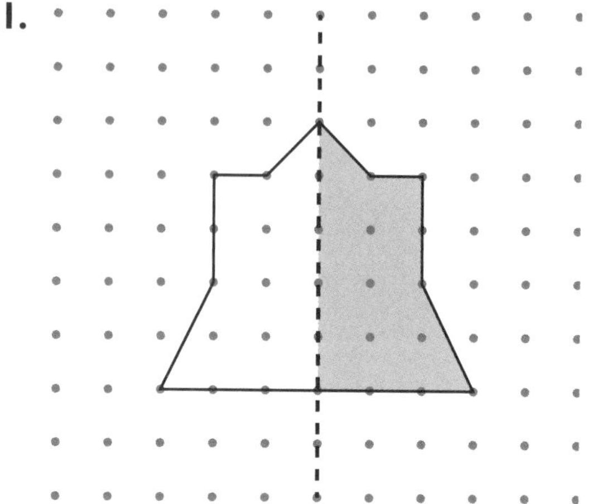

Reflective drawing shows understanding – accuracy may not be totally correct, but drawing shows concept. **(1 mark)**

22. 87 → 97

23 → 33

47 → 57

64 → 74

(1 mark)

23. 350 g **(1 mark)**

24. a)

b)

(2 marks: 1 mark for each correct clock)

25. 14 packets

Any viable method shown, e.g.

7 × 2 = 14

2 + 2 + 2 + 2 + 2 + 2 + 2 = 14

(2 marks for correct answer. Award 1 mark for using appropriate method but wrong answer given.)

26. 10 crayons. *Also accept 9 or 11.* **(1 mark)**

27. £1.50 **(1 mark)**

28. ○ △ △ **(1 mark)**

29. A 10 cm line drawn correctly with a ruler. **(1 mark)**

30. 81 —take away 10→ $\boxed{71}$

46 —take away 10→ $\boxed{36}$

97 —take away 10→ $\boxed{87}$

(1 mark)

Set C Paper 1

Practice question 6

1. 3 **(1 mark)**

2. 12 **(1 mark)**

3. 6 **(1 mark)**

4. 26 **(1 mark)**

5. 6 **(1 mark)**

6. 6 **(1 mark)**

7. 52 **(1 mark)**

8. 42 **(1 mark)**

9. 28 **(1 mark)**

10. 7 **(1 mark)**

11. 10 **(1 mark)**

12. $\boxed{55} - 12 = 43$

$43 + 12 = \boxed{55}$

(1 mark: both correct for 1 mark)

13. 5 **(1 mark)**

14. $24 \div \boxed{2} = 12$

$\boxed{12} \times 2 = 24$

(1 mark: both correct for 1 mark)

15. 12 **(1 mark)**

16. 25 **(1 mark)**

17. 24 **(1 mark)**

18. 6 **(1 mark)**

19. 61 **(1 mark)**

20. 25 **(1 mark)**

21. 30 **(1 mark)**

22. $20 \div \boxed{5} = 4$

$\boxed{5} \times 4 = 20$

(1 mark: both correct for 1 mark)

23. 31 **(1 mark)**

24. 33 **(1 mark)**

25. 5 **(1 mark)**

Set C Paper 2

Practice question 15

1. 50p 20p 2p
in any order **(1 mark)**

2. 41 **(1 mark)**

3. 11 children **(1 mark)**

4. 8 vertices **(1 mark)**

5. (centimetres) **(1 mark)**

6. 15 + 15 = 30 **(1 mark)**

7. 48 satsumas **(1 mark)**

8. 10

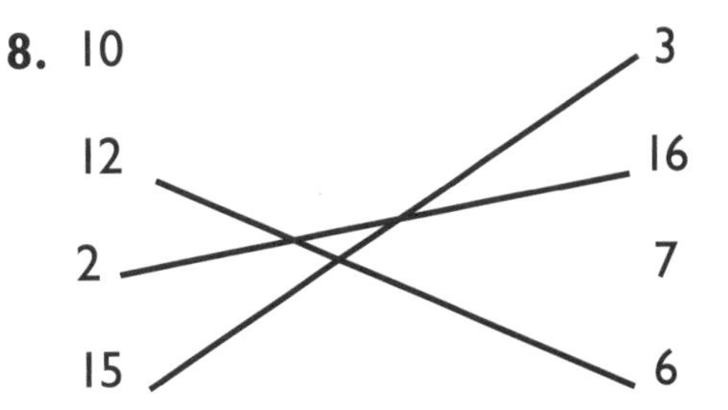

(1 mark)

9. 500 g **(1 mark)**

10. *Accept:*

10 lots of 5

5 + 5 + 5 + 5 + 5 + 5 + 5 + 5 + 5 + 5

10 × 5

or any drawn answer that clearly relates to 10 groups of 5. **(1 mark)**

11. a)

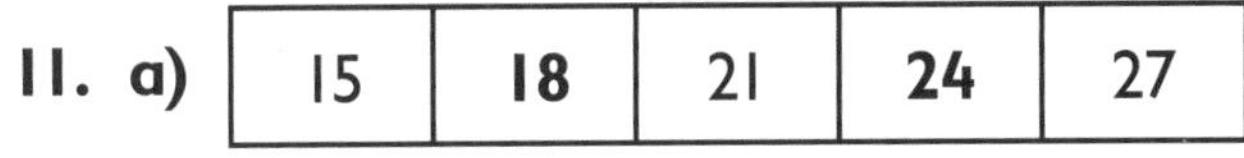

15	**18**	21	**24**	27

(1 mark)

b)

55	**50**	45	**40**	35

(1 mark)

12. $\frac{1}{3}$ *or* $\frac{2}{6}$ **(1 mark)**

13. (22°C) **(1 mark)**

14. a) 3 children **(1 mark)**

b) Light brown **(1 mark)**

15. (0°C) **(1 mark)**

16.

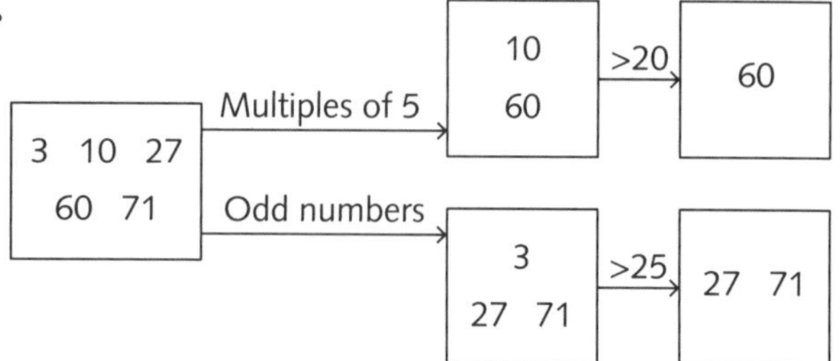

(2 marks: 1 mark for each correct row)

17. 45 minutes **(1 mark)**

18.

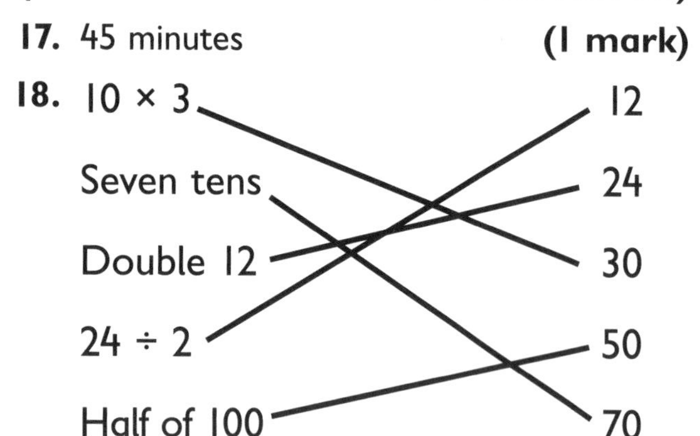

(1 mark)

19. 15 **(1 mark)**

20. **(1 mark)**

21.

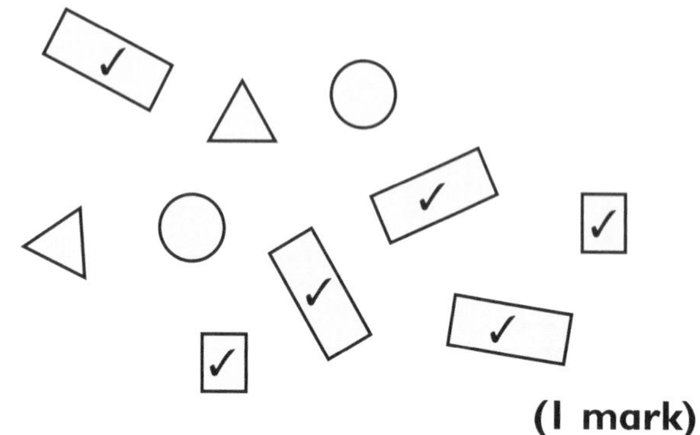

(1 mark)

22. *Any combination of available coins that total 87p, e.g. 50p, 20p, 10p, 5p, 2p.*

Do not *accept multiples of coins that are* **not** *available or amounts not represented by actual coins.* **(1 mark)**

23. 2 litres **(1 mark)**

24. These grid references shaded.

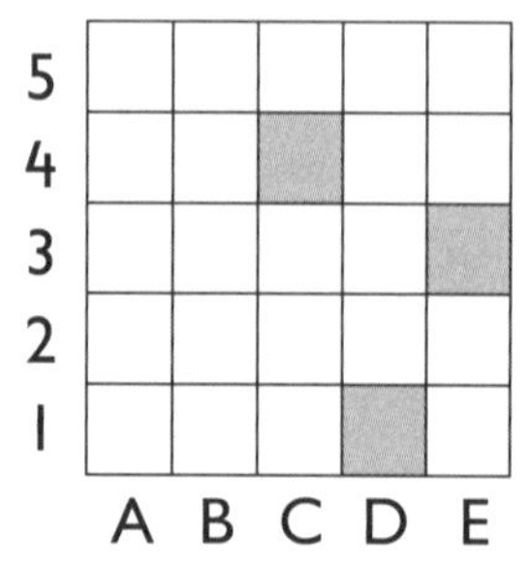

(1 mark)

25. 9 weeks **(1 mark)**

26. **a)** 8 × 10 = 94 − 14 **(1 mark)**

b) 50 ÷ 5 = 2 × 5 **(1 mark)**

27. 4 children **(1 mark)**

28. Any 4 extra squares shaded. **(1 mark)**

29.

(1 mark)

30. **a)**

Weather in March

Weather	Tally	Total
Sunny	𝍸	5
Showers	𝍸 III	8
Cloudy	𝍸 𝍸	10
Rain	𝍸	5
Snow	III	3

(1 mark)

b) Showers column shows 8 and Rain column shows 5:

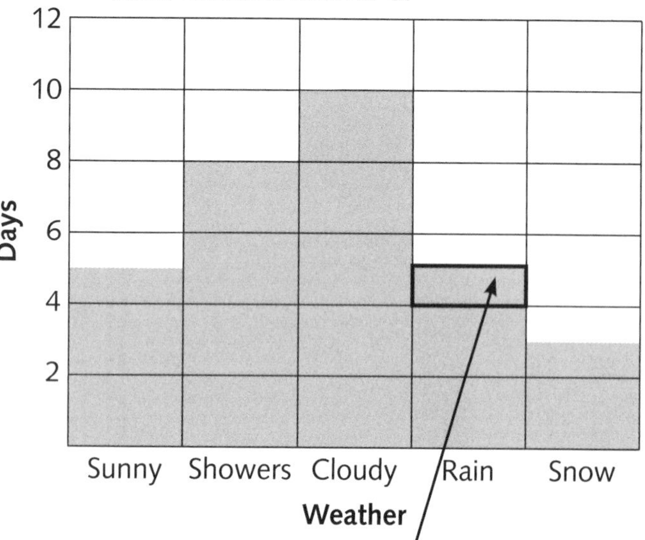

Note: It is important that the graph shows an understanding of each block representing 2 items and a split block counting as 1. **(1 mark)**

Aural questions administration

The instructions below are for the first five questions in each Paper 2.

Explain to your child that the first question is a practice question. When reading the question, remember to repeat the bold text in the question. Before proceeding to question 1, ensure that your child knows where they should have written their answer and discuss with them how they worked out their answer. Allow them to change their answer to the correct one by crossing out or rubbing out, to make sure they know the way to correct errors.

Before reading out questions 1 to 5, tell your child that they should try to answer all of the questions. If they can't answer a question, they should move on to the next one and come back to that one later. Remind your child that you can't help answer these next questions and that they should try to work them out on their own. Read questions 1 to 5, allowing time for your child to write their answers. When reading the question, remember to repeat the bold text in the question. At the end of each question, allow sufficient time for your child to complete what they can.

Set A Paper 2

Read out these questions to your child. Explain that they should listen carefully and write their answers on pages 18–19.

Practice question

This is a practice question for us to do together.

Amy has 12 caterpillars. **Look at the caterpillars. Circle half of the caterpillars.**

1 Question 1. **What number is 8 less than 27?** Write your answer in the box.

2 Question 2. **There are 5 packs of stickers. Each pack has 5 stickers. How many stickers are there altogether?** Write your answer in the box.

3 Question 3. Emma's baby sister has just been born. **Circle the weight that you think Emma's baby sister is most likely to be.**

4 Question 4. **A banana costs 15p. How much money would 3 bananas cost?** Write your answer in the box.

5 Question 5. Look at the shapes. **Two of the shapes are pentagons. Tick (✓) the shapes that are pentagons.**

Set B Paper 2

Read out these questions to your child. Explain that they should listen carefully and write their answers on pages 46–47.

Practice question

This is a practice question for us to do together.

One pack has 10 stickers. How many stickers would there be in 5 packs? Write your answer in the box.

1 Question 1. **Write the number three hundred and seven.** Write your answer in the box.

2 Question 2. **February is the second month of the year. What is the fifth month?** Write your answer in the box.

3 Question 3. **What is the difference between 58 and 45?** Write your answer in the box.

4 Question 4. **A cuboid has how many faces?** Write your answer in the box.

5 Question 5. **Alba watches her favourite film. How long will the film probably last?** Circle the time that is most likely.

Set C Paper 2

Read out these questions to your child. Explain that they should listen carefully and write their answers on pages 74–75.

Practice question

This is a practice question for us to do together.

Here is a number sequence. What number would be next in this sequence? Write your answer in the box.

1 Question 1. **Hannah has 3 coins which total 72p. Which 3 coins must Hannah have?** Write your answer in the boxes.

2 Question 2. **What is 12 less than 53?** Write your answer in the box.

3 Question 3. **There are 5 children on a bus. At the next stop 9 more children get on and 3 children get off.**

How many children are left on the bus? Write your answer in the box.

4 Question 4. **A cube has 6 faces. How many vertices does it have?** Write your answer in the box.

5 Question 5. **Max needs to know how tall he is. Which standard units will he use?** Circle the correct standard unit used to measure length and height.